Functions of
Biological Membranes

Other titles in the series

Cell Differentiation	J.M. Ashworth
Cellular Development	D.R. Garrod
Biochemical Genetics	R.A. Woods
Biochemical Pharmacology	B.A. Callingham

OUTLINE STUDIES IN BIOLOGY

Editors : Professor T.W. Goodwin, F.R.S., University of Liverpool
Dr J.M. Ashworth, University of Leicester

Editors' Foreword

The student of biological science in his final years as an undergraduate and his first years as a postgraduate is expected to gain some familiarity with current research at the frontiers of his discipline. New research work is published in a perplexing diversity of publications and is inevitably concerned with the minutiae of the subject. The sheer number of research journals and papers also causes confusion and difficulties of assimilation. Review articles usually presuppose a background knowledge of the field and are inevitably rather restricted in scope. There is thus the need for short but authoritative introductions to those areas of modern biological research which are either not dealt with in standard introductory textbooks or are not dealt with in sufficient detail to enable the student to go on from them to read scholarly reviews with profit. This series of books is designed to satisfy this need.

The authors have been asked to produce a brief outline of their subject assuming that their readers will have read and remembered much of a standard introductory textbook of biology. This outline then sets out to provide by building on this basis, the conceptual framework within which modern research work is progressing and aims to give the reader an indication of the problems, both conceptual and practical, which must be overcome if progress is to be maintained. We hope that students will go on to read the more detailed reviews and articles to which reference is made with a greater insight and understanding of how they fit into the overall scheme of modern research effort and may thus be helped to choose where to make their own contribution to this effort.

These books are guidebooks, not textbooks. Modern research pays scant regard for the academic divisions into which biological teaching and introductory textbooks must, to a certain extend, be divided. We have thus concentrated in this series on providing guides to those areas which fall between, or which involve, several different academic disciplines. It is here that the gap between the textbook and the research paper is widest and where the need for guidance is greatest. In so doing we hope to have extended or supplemented but not supplanted main texts, and to have given students assistance in seeing how modern biological research is progressing while at the same time providing a foundation for self help in the achievement of successful examination results.

Functions of Biological Membranes

M. Davies

Senior Lecturer in Biology,
University of York

Chapman and Hall

London

First published in 1973
by Chapman and Hall Ltd
11 New Fetter Lane, London EC4P 4EE
© 1973 M. Davies
Printed in Great Britain by
Fletcher & Son Ltd, Norwich

SBN 412 11350 3

Distributed in the U.S.A.
by Halsted Press, a Division
of John Wiley & Sons, Inc.
New York

Contents

1 Properties of the membrane phase

This book about the *function* of biological membranes starts with the assumption that the reader knows where membranes are found in living matter, has some idea of their thickness and appearance under the electron microscope after conventional fixation, staining, dehydration and embedding, and is wondering how the ubiquitous double track (dark, light, dark), and structures closely associated with it, fulfil the selective barrier properties so generally attributed to them. The philosophy which underlies this book is that if we want to understand biological membranes, we ought to think about, and experiment on them, as biochemists, yet be mindful whenever we can of two contexts, the conceptual framework of physicochemical principles which delimit that which is possible, and the physiological framework of discovered relationships which characterise that which has been sucessful in evolution. Section 1.1 is an attempt at a word picture of biological membranes as dynamic chemical structures; it illustrates how hard it can be to understand experimental results in these two contexts. The section is so tentative and impressionistic that it might seem more suitable to end with; the temerity of beginning with it can only be excused on grounds of the biochemical logic that inferences from chemical structure should be considered first whenever possible.

1.1 Membranes as dynamic structures.

Biological membrane fractions can be separated from other cellular matter by a variety of techniques with varying degrees of success. Membranes are found to consist of mainly lipid, protein, and some water. The lipid is a complex mixture in which phospholipid predominates. Phospholipids are amphipathic substances having (charged) polar heads and long hydrophobic tails. It has long been known that when spread on water, in which the polar heads dissolve, such substances form monomolecular layers. A monolayer can be held behind a movable dam, the force on which can be measured and related to the enclosed area. At large areas the film behaves as the two dimensional analogue of a gas; as the area is decreased a point is reached at which 'condensation' to a two dimensional 'liquid' occurs. For pure substances the area per molecule at this point has been convincingly related to the geometry and packing of the nonpolar chains. Langmuir discovered that such films could be folded over on themselves to form multi-layers (Fig. 1.1) in which non-polar and polar surfaces were alternately apposed. Amphipathic substances were found to emulsify mixtures of immiscible polar and non-polar liquids in the form of spherical droplets coated with monolayers; in aqueous solution spherical micelles (i.e. bilayer envelopes) could be found under certain conditions, more complex phases (liquid crystals) being formed as the proportion of water was decreased (Fig. 1.1). Gorter and Grendel calculated from monolayer studies and analysis of red cell lipids that there was just enough lipid in one cell to form a bilayer of surface area equal to that of the cell. Davson

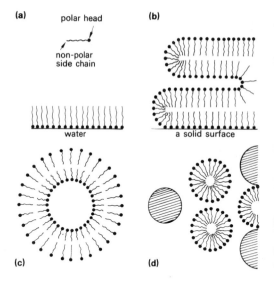

Fig. 1.1 Amphipathic molecules forming
(a) a monolayer
(b) multilayers
(c) bilayer envelopes
(d) liquid crystals – 'neat phase'.

and Danielli subsequently proposed their lipo-protein bilayer model for plasma membranes, in which protein coated both sides of a lipid bilayer. A variety of early evidence, including X-ray diffraction, indicated that myelin consisted of a multilayer of alternating lipid and protein. More recently, bilayers (and in part multilayers) separating two aqueous compartments have been made from synthetic lipids or extracts from membrane preparations; they mimic biological membranes well enough to have become indispensable research tools.

An obvious but perhaps superficial interpretation of much evidence from biological electron microscopy is that the stained double track corresponds with a bilayer having a plane of symmetry, a structure plausible enough in terms of the physicochemical background, but needing modification in order to explain the known

physiological properties. Doubts about the nature of the bilayer arise from the finding [29], by means of X-ray diffraction, of extensive dimensional and structural changes during conventional preparation for electron microscopy. Is the typical appearance after osmium staining evidence for a symmetrical structure? If the stain is specific for a particular component irrespective of its location, then yes; but if it stains non-specifically that which lies on the outside of an impenetrable layer, probably no. Surprisingly few experiments appear to have been directed at the penetration and specificity of stain. The problem of penetration of foreign molecules into the membrane structure is now being tackled in a more general fashion by observing, using various spectroscopic techniques, energetic changes in suitable 'probe' molecules due to their penetration into membranes and interaction there with surrounding molecules: this approach should yield more specific evidence about native membrane structure than conventional electron microscopy. Meanwhile it is physicochemically reasonable to accept the idea that a bilayer is the basis of membrane structure, while noting that the macroscopic study of monolayers and bilayers is no more likely to reveal physiologically important irregularities than the macroscopic study of crystals is to reveal packing defects. We may also suppose that the chance of finding by sectioning, and of resolving, distinctly stained macromolecular irregularities in a bilayer structure which might account for its physiological behaviour, would be remotely low.

The lipids of membranes may be subdivided into various categories which are summarized in Fig. 1.2 together with a diagrammatic indication of the different composition, on a dry weight basis, of four common membrane types. The proportion of those lipids which all membranes have in common varies widely; other lipids are restricted to specific types of membrane. Analysis into different subclasses of the phospholipids of human red cell extracts is shown

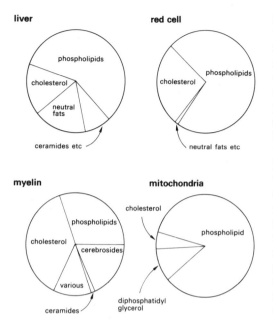

Fig. 1.2 Lipid composition of membranes

Table 1.1. Phospholipid components of human red cells [41].

Class	% (w/w) and S.E.	charge type
sphingomyelin	23·8 ±1·3	+ −
phosphatidyl choline	29·5 ±1·2	+ −
phosphatidyl ethanolamine	25·7 ±2·5	+ −
phosphatidyl serine	15·0 ±1·7	+ −
phosphatidyl inositol	2·2 ±1·0	−
lysolecithin	0−1·6	+ −
other	3·3 ±1·6	
recovery of P	95%	

in Table 1.1, together with the charges borne by the phospholipids at about pH 7. In man the rank order of phospholipid fractions in myelin is quite different from that in red cells. The various saturated and unsaturated fatty acid side chains are not distributed equally amongst the phosphoglyceride classes of red cells, neither are the side chains positioned randomly in an individual phosphoglyceride class, for instance, the 2 position of phosphatidyl choline tends to bear a poly-unsaturated fatty acid. In comparing these tissues we have seen no evidence for stoichiometric interaction between the different membrane components. This impression is confirmed by interspecies comparisons of a tissue, for example in mammalian erythrocytes the ratio of phospholipid to cholesterol is roughly constant, but there are marked differences in the rank order of different phospholipid fractions, the possible physiological significance of which is discussed below.

The amount of protein associated with lipid in a membrane preparation depends somewhat on the preparative biochemical procedure, but clearly varies from one membrane type to another, for example it is much lower in myelin than in red cell membranes. Membrane 'bound' proteins include enzymes, for example ATPase (in red cells and mitochondria) and dehydrogenases (in mitochondria and bacteria). In some cases it has been shown that extraction of lipid decreases the enzyme activity, but that it can be restored on adding back the appropriate lipid. It is conceivable that some proteins are attached to membranes solely by interaction with phospholipid head groups, but it is known that many membrane bound enzymes have significantly higher proportions of amino acids with apolar side chains than is typical of supernatant enzymes. It may well be that they have hydrophobic surface patches which penetrate and interact with the lipid part of the membrane. Bretscher [4] has found good chemical evidence (specific, but too complex to summarize here)

9

for the presence in each red cell membrane of about 5×10^5 copies of a protein of unknown function which spans the membrane, and may be assumed to have hydrophilic ends and a hydrophobic middle. This protein is probably identical with the spanning particles seen in electron micrographs of 'freeze cleaved' red cell membranes. Thus it is not too speculative to believe that proteins could exist which could traverse the membrane, presenting a hydrophilic patch to one side or the other.

Water is present in native membranes, and proton magnetic resonance studies [7] indicate that much of it is associated with protein. Most of the water is loosely bound, the molecules being considerably less mobile than in liquid water, and a little is tightly bound comparable with water in a solid hydrate.

Evidence of the dynamic state of membranes comes from radiotracer studies of the rate of renewal of membrane components. For example the turnover times of phospholipids [27] in mitochondria are expressed in weeks down to a couple of days, whereas for the myelin lipids they are expressed in months; furthermore there is no apparent correlation between the amount of a phospholipid fraction in the tissue and its turnover rate. Composition and turnover rates are tissue specific, yet both show a degree of independence. If, as is likely from their size, membrane proteins are each associated with more than one lipid molecule, then the variety of lipid turnover rates probably means that lipids and proteins change partners. A few more points about membrane dynamics may be made using the red cell as an example because of its relative simplicity and isolation. Quite extensive changes in the lipid composition of red cells can be brought about, or reversed, by dietary changes in times much less than the lifetime of an erythrocyte. A proportion of such a composition change appears to be brought about by acyl transferases associated with the membrane, but since the red cell contains no phospholipase, the precursors (e.g. lysophosphatide and fatty acid) must both come from the plasma. Since something must move for the substrates to get at the active site of the enzyme, the dependence on a membrane associated enzyme argues for the mobility of membrane components away from their postulated normal orientation in the bilayer. In man 60% of red cell phosphatidyl choline is exchangeable with that in plasma, and 30% of phosphatidyl serine. These differences in exchangeability cannot usefully be compared with differences of solvent extractability of phospholipids since that which is left after extraction is obviously not the whole membrane. Exchange of cholesterol between cells and plasma is complete in about 8 hours, whereas only about 10% of the phosphatidyl choline exchanges in 12 hours. Experiments on the reversible depletion of red cell cholesterol have also been contrived: surface area decreases and osmotic fragility increases (which might be due to permeability or mechanical changes or both). The association of changes in physiological properties with a change of lipid proportion brings us back to the notion that maintenance of particular ratios should be correlated with physiological parameters.

The reader must by now have noticed, even from this qualitative account, that the lack of comprehensible correlations of analytical data is a feature of the present topic: more extensive reviews [40] confirm the impression. To drive the point home, consider a simplified presentation of a species comparison for mammalian erythrocytes between a rather crudely determined physiological parameter, a permeability derived from the haemolysis time in ethylene glycol solutions, and two sets of composition data. Table 1.2 shows a reasonable correlation in terms of rank order. Looking *only* at this table one is tempted to conclude that the basis of the relationship could be that nearly all the saturated fatty acid is in the phosphatidyl choline fraction. In fact a glance at a table of fatty

Table 1.2. Species comparison of red cell properties.

phosphatidyl choline (%)	ox<sheep<pig<man≈rabbit≈rat
glycol permeability	sheep<ox<pig<man<rabbit<rat
saturated fatty acid (%)	sheep<ox<pig<rabbit<man<rat

acid distribution shows just how wrong this would be: one is evidently examining but one facet of a multifactored correlation, and the mechanism which underlies it is not understood.

In living matter biological membranes separate aqueous regions of different composition, and it can be seen from light microscopy of epithelial cells, stained for carbohydrate and protein, that the difference of composition extends to the cell boundary, on the outside of which carbohydrate predominates and on the inside, protein. By implication might one expect that a membrane of mixed composition would respond asymmetrically to the molecular imbalance of its surroundings? Chemical evidence [5] for asymmetry has been found from the reaction of intact erythrocytes with group-specific impermeant reagents as contrasted with the reaction of osmotically lysed, and washed, erythrocyte membranes (ghosts) both the inside and outside of which are accessible. For example, many more proteins are associated with the inside of the red cell membrane than the outside, and there is much less of reactive phosphatidyl serine and choline outside than inside. Much evidence shows that groups concerned with cell recognition are on the outside of the membrane.

In pure synthetic lipids, at biological temperatures, the hydrocarbon tails of the molecules may pack in orderly fashion, like crystalline solids, but thermochemical studies of natural lipids show that the mixed fatty acid side chains have the irregularity and mobility associated with a liquid. Evidence for the mobility of the lipid in the membrane has been sought from experiments on growing *Escherichia coli* [12, 28],

which show that newly synthesized lipid mixes with pre-existing lipid in times short compared with the mean generation time, but which at present do not distinguish between the possibilities of random insertion throughout the membrane or insertion at a few centres followed by lateral diffusion. It has been shown by spectroscopic studies of model bilayer systems that lateral diffusion (in the plane of membrane) is much quicker than the change of place ('flip-flop') between one side of the membrane and the other. If these findings can be related to growing cells, they help to clarify how asymmetry of membranes could be maintained in a dynamic steady state.

Summary
This sketch leads one to think of an asymmetric, two dimensional liquid structure of randomly associating molecules with different lifetimes in the membrane, the physiologically appropriate proportions being under ultimate genetic control, often apparent in terms of tissue and species differences yet capable of modification by environmental factors as if genetic control were exerted indirectly by the levels of specific biosynthetic and catabolic enzymes. The specific association of protein with lipid probably depends on the formation of apposite clusters of lipid molecules which are neither stationary nor permanent. Assembly of the membrane does not require enzymes: thermodynamically it has to be where it is. This dynamic picture serves to blur the static image of the basic bilayer which we tend to derive from cytological studies.

1.2 Diffusion, and osmosis

1.2.1 Diffusion
The diffusion of a substance in solution can conveniently be studied theoretically and practically when a concentration gradient exists along one defined axis. Empirically, this diffusion is described by Fick's first law:

11

$$\frac{dm}{dt} = -DA\left(\frac{dc}{dx}\right) \qquad 1.1$$

where dm/dt is the flow of matter (in mass or moles per time) normal to a plane of area A, at which the concentration gradient is dc/dx (where c is measured in units consistent with those of m). D, the diffusion coefficient, is a small number in the usual units ($cm^2.s^{-1}$) and in aqueous solution it does not change rapidly with molecular weight, M.

Table 1.3. Diffusion coefficients in aqueous solution at $25°C$ [10, 16].

Substance	O_2	acetyl choline	sucrose	serum albumin
D cm^2.s^{-1} ($\times 10^6$)	19·8	5·6	2·4	0·7
M	32	182	342	69×10^3

The classical interpretation of diffusion in terms of molecules changing place with one another in solution as a result of random thermal motion, is that of Einstein. He derived two important equations which are still used in the interpretation of diffusion measurements. The first (Equation 1.2) shows that the diffusion coefficient for a specified molecular species varies in

$$D = \frac{RT}{L}\left(\frac{1}{f}\right) \qquad \text{(where } f = 6\pi\eta r) \qquad 1.2$$

different solvents and is inversely proportional to the frictional coefficient f between solute and solvent (R is the gas constant, T the absolute temperature, and L the Avogadro number). Assuming spherical solute molecules of radius r in a medium of viscosity η, in which the solute obeys Stokes' Law, $f = 6\pi\eta r$, the diffusion coefficient and molecular size can be related theoretically, though it appears that in practice the assumptions are imperfectly valid. We can see from Equation 1.2 and Table 1.3 that, as a transport mechanism, diffusion in free solution

is not very selective with respect to molecular size; the values of D are roughly proportional to $M^{-1/2}$ for small molecules and $M^{-1/3}$ for larger ones. Perhaps more important from the point of view of membrane transport is the comparison between diffusion in water and oil; taking vegetable oils as roughly analagous media to biological membranes, we conclude from viscosities that D for the same substance in oil would be 10^{-2} fold that in water, or less.

Einstein's second equation, 1.3 is important in biology; it is used to calculate Δ, the distance

$$\Delta^2 = 2D\tau \qquad 1.3$$

moved by a molecule in a very short time τ, during which dc/dx is assumed to be effectively unaltered. In this form the equation applies to a situation in which the concentration gradient is unidimensional. (The reader may use Equation 1.3 to calculate the time for a 20 nm trans-synaptic journey of acetyl choline, and compare it with the synaptic delay measured by neurophysiologists). Knowledge of the time taken for molecules to diffuse within cells is relevant to studies of biological transport. For diffusion in three dimensions Equation 1.4 replaces Equation 1.3.

$$\Delta^2 = 6D\tau \qquad 1.4$$

Suppose a molecule is produced biosynthetically at the centre of a cell (say a spherical bacterium of radius $0·7\mu m$): how long does it take to diffuse to the cell membrane?

We can calculate this for a molecule the size of sucrose, using the value of D, $10^{-6}cm^2s^{-1}$, estimated for sucrose in the exudate of broken bacterial cells [25]; the time is of the order 10^{-3}s. The significance of this calculation may be seen by taking another kind of example. There is evidence from tracer labelling experiments in bacteria that many different kinds of small molecules behave as if in well mixed single 'compartments'. Many such molecules 'turn over' [11] very rapidly, for instance the average ATP molecule in *E. coli* is renewed every second. Such rapid turnover must depend on good mixing, so we might guess that τ for diffusion from the centre to the periphery of the cell should be 1/100 the turnover time or less ($\tau \leqslant 0\cdot01$ s). For a turnover time of one second it follows that $D \geqslant 0\cdot1 \times 10^{-6}\,cm^2.s^{-1}$. Such a range of D means that molecules up to the size of large proteins would behave as if in single compartments, if their medium were water. In the bacteria this could hardly be true for the proteins, because of their entanglement with larger obstacles such as ribosomes and DNA, but it is reasonable for smaller molecules which will move in water filled spaces in a medium more fluid than the macroscopic viscosity of the cell contents would indicate. This argument is justified by the fact that the value of D for sucrose in the bacterial exudate was as much as 40% of D for sucrose in pure aqueous solution.

It is interesting to consider the case of a spherical cell only 5 times the diameter of this hypothetical bacterium (and many mammalian cells are bigger still). For the chosen criterion of mixing ($\tau \leqslant 0\cdot01$ s, centre to periphery) we find $D \geqslant 2\cdot5 \times 10^{-6}$, which would only be true for very small molecules such as oxygen. This calculation prompts the teleogical comment that the subdivision of eukaryotic cells into compartments bounded by membranes may have arisen partly in order to take advantage of the random thermal motion of small molecules to ensure that reactions would go at steady controllable rates.

1.2.2 Osmosis

Physical chemists often think of the osmotic pressure of a solution as a thermodynamic function whereas it is more useful for biologists to consider osmotic pressure as an operationally defined quantity, which depends both on the solution and a membrane which is selectively permeable to one or more solution component. In the familiar elementary example the single solute B is dissolved in the single solvent A, and the semipermeable membrane is a molecular sieve which allows A but not B to pass. The osmotic pressure of the solution is defined as that pressure which must be exerted across the semipermeable membrane separating solution from solvent, in order to prevent the flow of A. By induction from experimental observations for dilute solutions the limiting law $\pi = RTc$ was obtained, where c is the molar concentration of the solute B. The biologist who is particularly concerned with the mechanism of permeation has to study osmotic pressure in a more complicated situation. Although it was apparent in Section 1.1 that permeation of membranes by substances dissolving in them may be a better model for biological membranes than sieving by size and charge, it has been much less studied. Experimental and theoretical study of molecular sieves has enabled biologists to understand what happens if the membrane is not truly semipermeable, and B leaks through, albeit slowly compared with A. In such cases the apparent osmotic pressure π' may be defined as that pressure difference which must be exerted across the membrane in order that there shall be no volume flow through the membrane. The wide availability of artificial membranes for technological applications, e.g. cellulose derivatives with attached groups in the case of ion exchangers, has resulted in many biophysical model studies of molecular sieves. Often the interpretations have been complicated and disappointing in their lack of relevance to biology, but some useful general principles have been established.

Let us start the summary of these principles with a fundamental contrast: in an osmometer with a semipermeable membrane, a stable equilibrium can be set up by exerting a hydrostatic pressure difference p across the membrane given by $p = RT\Delta c$, whereas with a solute-permeable membrane the only possible equilibrium (at 'infinite time') is that at which $p = 0$ and the solute has diffused through the membrane until $\Delta c = 0$, so that no osmotic difference remains. Apparently instantaneous values of π' can be determined at intervals after setting up the osmometer. They decrease with time and, even if extrapolated back to zero time, are all less than the theoretical value calculated from the Van't Hoff equation. If a series of chemically homologous polymers of different molecular weight is compared at constant Δc with the same membrane, it is found that π'/π decreases with decreasing molecular size from a maximum of 1 for any impermeant polymers (with respect to which the membrane is semipermeable). The ratio π'/π is called the reflection coefficient (symbol σ) and has been determined for a range of biologically interesting molecules both for artificial membranes and, provided only passive transport was concerned, for cell membranes.

The observations can be explained qualitatively using a simple kinetic approach. If solution, on the $''$ side, is separated from solvent ($'$ side) by an impermeable flexible membrane, the hydrostatic pressure equalizes across the membrane, and if χ stands for the bombardment pressure on the membrane due to the molecules, specified by the subscript (A is solvent and B is solute), then

$$\chi_A'' + \chi_B'' = \chi_A',$$

and therefore the difference of bombardment pressure due to solvent is equal to the solute bombardment pressure:

$$\chi_A' - \chi_A'' = \chi_B'' \qquad 1.5$$

If, at an instant, the membrane is suddenly made semipermeable this difference of solvent bombardment pressure is uncompensated, and the left of this equation is the pressure which results in an osmotic flow; it is therefore the pressure which must be applied to prevent the flow, and is proportional to the solute bombardment pressure. On the other hand if the membrane were instantaneously made leaky to B as well,

the left of Equation 1.5 would not represent the pressure resulting in osmotic flow. If $(1 - \sigma)$ is defined as the proportion of solute molecules which, at the instant considered, happen to have such an orientation and rate of approach to the molecular sieve that they pass through, 1.6 is the equation of which the left side represents the pressure resulting in osmotic flow;

$$\chi_A' - \chi_A'' - (1 - \sigma)\chi_B'' = \sigma\chi_B'', \qquad 1.6$$

the right side is due to the proportion of solute molecules reflected back by the membrane. At this instant, therefore, the pressure π' which must be applied to stop the flow is proportional to $\sigma\chi_B''$. The ratio π'/π is therefore equal to σ, the reflection coefficient. The reader should not conclude from this explanation using a molecular sieve as an example that there are fixed and static apertures in the membrane: elements of the intermeshed polymeric molecules of the membrane take part in thermal motion, and the permeation of solvent or solute should be thought of as occurring through mobile apertures.

A final contrast concerns solvent flow in osmosis and free diffusion. Though both depend on random thermal motion. the characteristics of these flows differ: whereas in free diffusion, solvent and solute change places randomly, in osmosis the solvent flow is streamlined (with just a slight thermal wobble). It has been shown fairly recently that the rate of flow of A (in the absence of B) through a given membrane under a hydrostatic pressure difference p is equal to the rate of flow of A in an osmometer through the same membrane at a concentration of B such that the osmotic pressure π equals p in the previous case.

1.3 Permeability

For a long time the properties of the membrane phase have been explored, in a different way from those mentioned in Section 1.1, by measuring the rates of penetration of non-metabolized (and usually non-natural) molecules into cells. These molecules are thought of as probes of membrane structure in that correlations are sought between their physical characteristics and their penetration rates, and that hypotheses of membrane structure are devised with a view to explaining the correlations. Before summarizing some findings, the common terminology used in these studies is reviewed as an aid to understanding the original literature.

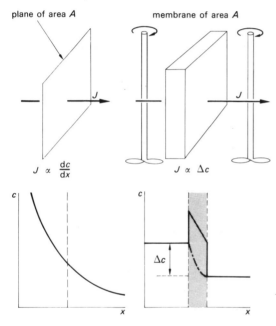

Fig. 1.3 Contrast between diffusion in free solution (left) and through membrane (right) between stirred compartments. Corresponding graphs show concentration c against distance x; graph for membrane shows two different hypothetical concentration profiles.

The word *flux* is used in the quantitation of flows, for instance (Fig. 1.3) when a physical chemist studies diffusion down a continuous unidimensional concentration gradient, which is measurable by physical techniques, the flux J is the rate of flow of matter per unit area. If a plane of area A is considered, $J = 1/A \, (dm/dt)$, where m is usually in moles (see Equation 1.1). If penetration through a membrane occurs by diffusion obeying Fick's law, the same equations apply to the flux through a planar membrane separating well stirred compartments, but the concentration gradient is now inside the membrane and almost invariably hidden from measurement. The concentration difference Δc can be measured, and Fick's law becomes $J \propto \Delta c$. When a relationship of this form is found, the complete equation for J includes a *permeability coefficient* (sometimes called a constant). There are many alternative definitions of such coefficients, according to the experimental context and the units of measurement; the interconversion of units for solute permeability has been summarized by Stein. Biological flows are often expressed as fluxes, but usually the relevant area cannot be measured, and the fluxes are referred to unit weight of cells or other similar measures. Under constant experimental conditions the area may be assumed proportional to the measure used, but it may be unwise to assume that equal fluxes under different physiological conditions refer to equal areas.

If, for present purposes, c_0 is the outside concentration and c_i the inside concentration, and P is the permeability coefficient for a solute, then its net flux J in the direction outside to inside is given by Equation 1.7. The reader should check the dimensions of P, given that

$$J = P(c_0 - c_i) \qquad 1.7$$

those of J are mass/length^2 × time. Although diffusion down a concentration gradient is implied when Equation 1.7 is obeyed in a biological system, the gradient is inaccessible to measurement, and the form of the concentration profile cannot even be deduced because we have reason to believe that the molecular properties of the membrane are not isotropic.

15

Biologists often speak and write loosely about concentration gradients when all that they can measure are concentration differences. The operative difference is actually that between two well stirred compartments. In studies on single cells, or sheets or tubules composed of layers of cells, the outside compartment can be mechanically stirred, and indeed must be in many cases to ensure adequate oxygenation; the cell interior is stirred by diffusion, and also by the active motion of replicating DNA (in bacteria), ribosomes and probably mitochondria. We know from experiments that in some cases neither of these agencies is adequate to eliminate significant unstirred layers adjacent to the membrane. Diffusion of solute through an unstirred layer contributes to the permeability coefficient calculated on the basis of the overall concentration difference; putting it in another way, the true concentration difference across the membrane is less than that used to calculate P. It is therefore a technical matter of some importance to reduce unstirred layer effects to negligible proportions, or to devise means of correcting for them.

Measurements of permeation need not be confined to solutes; water permeability may be measured, either in a steady state by studying the flow of tracer (D_2O or THO) after its addition to one side of the system, or by contriving an osmotic difference as a result of which water flows. A word of caution about the significance of water permeability results is necessary: it has been shown that water and solute permeability are interdependent, as the reader may indeed have recognized from the treatment of diffusion and osmosis in the previous section. To cope with this interdependence parameters for permeation based on the thermodynamics of irreversible processes are now coming into use. The only one of these parameters which we need to use for the biological examples chosen for this book is the reflection coefficient σ, introduced in the previous section.

A range of non-electrolytes which are not metabolized has often been used in permeability studies for example urea, alkyl-ureas, propionamide, trimethyl citrate and ethylene glycol. In the work reviewed in this section, membrane permeation follows a Fick's law relation, and the reader may also assume that when equilibrium is reached, the concentration of permeant is equal on both sides of the membrane. The partition coefficient of the substance between a non-polar liquid (olive oil or ether) and water serves as a parameter of the polar or non-polar character of the probe molecule, and the molecular weight M is often a satisfactory parameter of size. The giant algae *Chara* and *Nitella* have been much used for permeability studies because of their experimental convenience. A correlation by Collander, taken from his classical systematic study of the permeability of *Chara*, is often reproduced: it is a log–log plot of $PM^{1/2}$ (i.e. the permeability coefficient corrected for the expected effect of molecular size on diffusion) against the olive oil/water partition coefficient. The points are fairly scattered, that for water being particularly anomalous, but the essential linearity indicates the important relationship between solubility in a bulk lipid phase and membrane permeation. The results have been considered further by Stein, who has plotted $\log PM^{1/2}$ against N, the number of hydrogen bonds the probe molecule makes with water: $\log PM^{1/2}$ decreases linearly as N increases from 2 to 8 per molecule. This correlation, which is less scattered than Collander's shows that the rate of permeation of the probe molecule depends on the energy required to break all, or a constant proportion, of the hydrogen bonds between the probe molecule and water. Unfortunately the full significance of this escapes us at present because we have no way of telling whether the rate determining step is diffusion through the membrane interior (in which case the concentration of the substance just within the edge of the membrane is the equilibrium

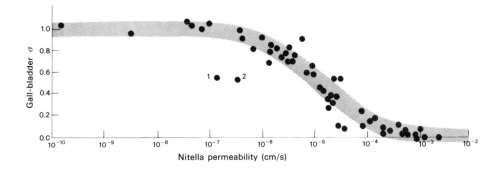

Fig. 1.4 Gall bladder σ against permeability of *Nitella* for the same compounds. Deviant points are 1, urea and 2, methyl urea. By courtesy of Dr. J.M. Diamond and Annual Reviews Inc. (see Bibliography).

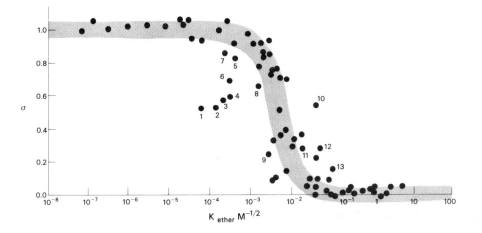

Fig. 1.5 Gall bladder σ against ether/water partition coefficient K with correction for dependence on molecular weight M. By courtesy of Dr. J.M. Diamond and Annual Reviews Inc. (see Bibliography).

concentration predicted from the partition co-efficient) or passage from water into the membrane (which would be expected to depend mainly on the forces between solute and water on which the partition coefficient also depends).

The probability that these correlations in the algae are of general significance has been strikingly confirmed by comparison with permeabilities in rabbit gall bladder determined by Wright and Diamond. A completely different method was used, the very rapid measurement of streaming potentials, from which σ could be determined for large numbers of compounds using a single preparation. Gall bladder σ values

(Fig. 1.4) for 52 compounds fell smoothly from 1 to 0, with only two exceptions, as permeability of *Nitella* to the same compounds increased.

Fig. 1.5 shows a relationship of similar form between the gall bladder σ and the ether/water partition coefficient divided by $\sqrt{M}$. Together with Fig. 1.4, this indicates the similar dependence of P and σ on molecular size. The pattern of non-electrolyte selectivity is therefore encouragingly similar in two very different biological systems.

The permeability of biological membranes to water is typically two or three orders of magnitude greater than to the small non-electrolytes such as glycerol or urea, though relative to the rate of diffusion of water in aqueous solution, water permeabilities justify our continuing to regard biological membranes as very effective waterproofing layers.

Permeabilities of biological membranes to water and some very small hydrophilic compounds have been measured with considerable care, and on various grounds, (too complex to summarize here) the results justify the retention of the notion that in addition to permeation through the lipid bilayer, these molecules also permeate through very small water-filled pores. However, it must be emphasized that the interpretation in terms of pores does not follow uniquely from the data, in fact the theoretical treatment remains to be justified, and some other hypothesis more in keeping with what we know of the dynamic nature of the cell membrane might well account for the results.

Bibliography

Jamieson, G.A. and Greenwalt, T.J. (eds) (1969) *The red cell membrane.* Lippincott, USA
Conference report, including chapters by R.S. Weinstein, on electron microscopy using the 'freeze cleavage' technique, and D.J. Hanahan on membrane composition.

Siekevitz, P. (1972) *Ann.Rev.Physiol.* **34**, 117.
A stimulating short review, entitled 'Biological membranes: the dynamics of their organization'.

Einstein, A. (1908, reprint 1956) *Investigations of the theory of the Brownian motion.* Dover, USA 68–85.
Very readable classic on diffusion.

Stein, W.D. (1967) *The movement of molecules across cell membranes.* Academic Press, New York.
A scholarly, integrated and original account of biological transport. Chapters 2 and 3 are about diffusion and permeability. Other chapters also relevant.

Diamond, J. and Wright, E.M. (1969). *Ann.Rev. Physiol.* **31**, 581.
A comprehensive review of electrolyte selectivity (see Section 3.4 of this book) and non-electrolyte selectivity by biological membranes.

Kedem, O. and Katchalsky, A. (1958), *Biochim. Biophys. Acta.* **27**, 229.
The pioneer review of the application to membrane transport of the methods of 'irreversible thermodynamics'.

2 Experiments pose problems

2.1 Passive transport

The probe molecules for which the results of permeation experiments were reviewed in the last section passed into the cell by Fickian diffusion: nearly all of them were non-natural compounds. This section is about the large class of probe molecules, natural compounds or very close relatives of natural compounds, whose permeation has shown departures from Fick's law, usually in experiments in which the compounds were not metabolized. However these permeants have a feature in common with those discussed in Section 1.3: when the permeants have come to equilibrium with the cells, the concentrations inside and outside are equal. This means that although Fick's law is often not obeyed by these compounds, their rate of permeation depends on some function of the external and internal concentrations which reduces to zero when these concentrations are equal. Evidently the cells lack any mechanism for accumulating these substances in the cells at concentrations greater than in the surrounding medium. On the basis of this criterion the transport of both groups of molecules, whether or not in accordance with Fick's law, is described as *passive*; this expresses the idea that the energy for their transport is a function only of the concentration gradient, and requires no special link with metabolism (compare Section 3.2). There is evidence that some compounds which are metabolized are also passively transported, for example glucose into erythrocytes, but the direct criterion of equality of concentration at equilibrium is inapplicable.

The entry of sugars into red cells has been surveyed comprehensively [24]. Many experiments were done using a rather indirect technique, the principle being that when the sugar gets into the cell it increases the effective osmotic pressure of the contents, water enters, and the cell swells. The scattering of light by a suspension of cells is sensitive to cell volume (decreasing as the cells swell, all else being equal) so that relative initial rates of permeation of the sugars may be calculated at different concentrations. The method can be calibrated so as to give absolute figures for the fluxes. More specific information, obtained with radioactively labelled sugars, confirms the general picture from the light scattering technique.

Double reciprocal plots of initial rates of entry J of sugar into previously washed red cells, against medium sugar concentration $[S]$ were straight lines, showing that sugar entry follows Michaelis Menten kinetics (Equation 2.1) rather than Fick's law. A few values at 37°C of the

$$J = \frac{[S]\,V_{max}}{K_m + [S]} \qquad (2.1)$$

Michaelis constant K_m and the maximum velocity V_{max} (per litre of cells) are shown in Table 2.1; they are abstracted from an extensive compilation by Stein. These findings indicate that sugar permeation depends on the presence of saturable entry mechanisms which have some of the characteristics of enzymes. If, in the absence of detailed knowledge of the mechanism and bearing in mind the usual warnings [11], the

Table 2.1. K_m and V_{max} for sugar transport in human red cells. 37°C.

Sugar	K_m(mM)	V_{max}(mM min^{-1})
glucose	4·0	328
galactose	39, 61	710
mannose	13	350
xylose	50, 71	650
ribose	2500	770
L.arabinose	220, 250	710

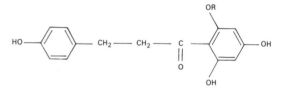

Fig. 2.1 phlorizin $R = C_6H_{11}O_5$; phloretin, $R =$ H; phlorizin is a glucoside; phloretin its aglycone

K_m is taken as an indication of the affinity of the transport mechanism for the sugar, the table shows much more variation in the affinity than the rate, as if the same entry mechanism served the different sugars. This point of view is enhanced by competition studies using two sugars, which show mutual inhibition of entry; in certain cases which have been studied quantitatively, it has been shown that the inhibitor constant K_i defined by Equation (2.2) in which $[I]$ is the concentration of the inhibitory sugar, is very roughly equal to the K_m for this sugar found in measurements of its transport.

$$J = \frac{[S]V_{max}}{(1 + [I]/K_i)K_m + [S]} \quad (2.2)$$

The glucoside phlorizin (Fig. 2.1), a competitive inhibitor of a number of enzymes with carbohydrate substrates, is a competitive inhibitor of sugar transport. Interestingly its aglycone, phloretin, is a much stronger inhibitor, and some related compounds with alkyl substituents are stronger still. This suggests that some non-polar region with an affinity for phloretin is contiguous with or part of the transport site which interacts with sugars.

The specificity of the sugar transport system is evidently rather broad, being comparable with that of the hexokinases. LeFevre has convincingly shown that the transport system prefers those sugars which have predominantly equatorial hydroxyl groups and also are more stable in the conformation shown in Fig. 2.2 than in

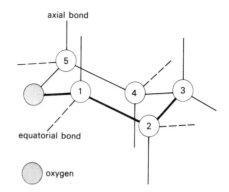

Fig. 2.2 The chair conformation preferred by the sugar transport site of red cells. In glucose all hydroxyls are equatorial.

the other chair conformation (formed by bending the oxygen up to make the chair back, and the 3 carbon down).

A number of thiol reagents are effective non-competitive inhibitors of sugar transport: they may be displaced from the red cell not by raising the sugar concentration, but by increasing the concentration of a thiol added to the medium. The many thiol groups on the cell surface differ in reactivity, for example at a low concentration of p-chloromercuribenzoate sulphonate, which inhibits sugar transport, only 0·6% of the thiol groups of the erythrocyte react. This leads to a maximum estimate of 7×10^5 sites per cell, which is comparable with an estimate of

2×10^5 per cell from the binding of ^{14}C glucose (at low medium concentration) to red cell ghosts. Expressing the concentrations of these sites on the basis of total cell volume, they come in the micromolar range, which is high compared with the concentrations of some enzymes.

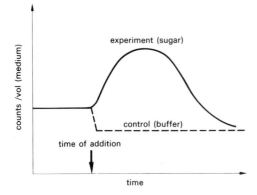

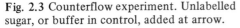

Fig. 2.3 Counterflow experiment. Unlabelled sugar, or buffer in control, added at arrow.

Rosenberg and Wilbrandt [33] drew attention to a most significant consequence of competition in sugar transport, the phenomenon of *counterflow*. In a typical experiment cells were first equilibrated with a labelled non-metabolized sugar at low concentration. When another sugar was added to the medium, unlabelled and at a higher concentration, it entered the cell but at the same time some of the labelled sugar came out, moving against its concentration difference. This transient concentration difference decreased until the concentration of the labelled sugar became equal inside and outside. Fig. 2.3 shows the type of result, and includes the graph of a control experiment in which a volume of buffer solution is added equal to the volume of un-labelled sugar solution in the first experiment; this shows clearly that the effect is not due to the inevitable dilution of the labelled sugar in the medium. This result implies that the influx

of one sugar forces a coupled exit of its competitor, the entry and exit mechanisms being linked. A related observation is that certain non-penetrant competitive inhibitors of entry, simultaneously inhibit exit of a labelled sugar. These results strongly suggest, though they do not prove, that the transport site is mobile and has access to the inside of the membrane where sugars may also compete. This model is discussed in detail in Section 3.1.

Experiments on sugars have been chosen as good illustrations of the phenomenon of passive transport. Other non-electrolytes, for example amino acids, are similarly transported in red cells, and passive transport systems with similar characteristics have been found in a variety of cells and tissues.

2.2 Active transport
The major proportion of all biological transport systems which have been studied, differ from those described in the last section in one important respect; they can function concentratively. Glucose, a ubiquitious source of carbon for cell growth and metabolism, is a very important substance to conserve. It has long been realized that in animals with kidneys the glomerular filtrate contains glucose at approximately the plasma concentration, whereas the urine seldom contains any; from this information alone we deduce that glucose must be reabsorbed against its overall concentration difference. In fact micropuncture experiments on the amphibian nephron showed that the glucose concentration fell to zero by the end of the proximal tubule. Numerous other substances are also reabsorbed in similar fashion in the kidney, this type of concentrative transport being known as *active* transport. In the kidney there is an upper limit to the rate at which glucose can be reabsorbed. Other sugars infused with glucose are also reabsorbed, the total rate of sugar reabsorption being equal to the maximum rate for reabsorption of glucose by itself. This is evidence of competition

21

for sugar transport similar to that observed in red cells. The active reabsorption of glucose is completely blocked by the inhibitor phlorizin (see Fig. 2.1). The widespread importance of active transport is evident from the briefest consideration of our physiological knowledge. Intestinal absorption of glucose would be extremely wasteful unless it were active, since passive transport would require the intestinal glucose concentration to be at all times greater than that of the plasma. Active transport was demonstrated *in vivo* in the nineteenth century when a sample of the animal's own plasma, introduced into a loop of the intestine was seen to be reabsorbed. In 1892 Reid [31] initiated the *in vitro* study of the absorption of water, by showing that water could be absorbed from the lumen of an isolated section of the intestine, even against a hydrostatic pressure difference. The study of intestinal absorption in isolated preparations continues to this day, the relationship between the metabolism of the mucosa, and the active uptake of amino acids and sugars from the lumen to a higher concentration on the serosal side, being a matter of continuing interest. A moment's thought provides numerous examples of the importance of active uptake of salts in the lives of freshwater animals.

While nearly all the examples of active transport in complex tissues have commanded the attention of experimentalists, often because of their medical interest, there is little doubt that fundamental understanding of the processes has been gained most readily from the more experimentally convenient systems, such as the simpler epithelial layers and single cell systems.

When *E. coli* is grown to stationary phase in a minimal medium of typical composition, containing a low concentration of K^+ (e.g. 10 mM), analysis [34] of cells collected on filters shows them to have a high content of Na^+ and a low content of K^+ comparable with the medium. When the cells are resuspended in fresh medium, growth recommences. In a period less than the

22

mean generation time the cells lose sodium, and gain potassium to an intracellular concentration about twenty times that of the medium. Similar observations were made in connection with the cold storage of blood for subsequent use in transfusions. Fresh red cells contain K^+ at a concentration of about 150 mM in the cell water, and Na^+ at roughly one tenth this concentration. In cold stored red cells the concentration of Na^+ is elevated and that of K^+ is reduced. On resuspending these cells in physiological saline at 37° containing glucose, the original concentrations were restored in about an hour. It was

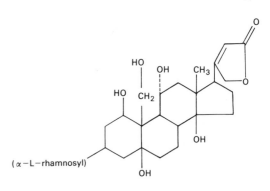

Fig. 2.4 Ouabain (strophanthin G)

found that the extrusion of three sodium ions was associated with the uptake of roughly two potassium ions, their transport was dependent on the presence of glucose, and the cardiac glycoside ouabain (Fig. 2.4) or its aglycone completely inhibited the uptake and extrusion processes. Ouabain does not arrest glycolysis, indeed the inhibition of ion transport in red cells by cardiac glycosides has been shown to have an ATP sparing effect. The evident analytical advantages of working with red cells were enhanced when it was discovered that washed red cell ghosts could be resealed by incubation at 37° in saline at physiological osmolarity containing Mg^{++} and ATP. The proportions of other inorganic cations in the saline could be adjusted by the experimenter. Using this technique it was shown the K^+ uptake by the ghosts increased with the internal Na^+ concentration, and there was no K^+ uptake without internal Na^+. There is an instructive difference between this experiment and that in *E. coli*; there is no detectable sodium requirement for the growth of *E. coli*, but potassium is essential, being taken up by growing cells from a sodium free medium.

It could be argued that in the above mentioned experiments on ion transport, cation movement was some kind of maintained counterflow secondary to the undetected movement of some other substance. The classical demonstration, by Ussing and Zerahn of active transport in frog skin, disposes of this argument, and introduces discussion of a definition of active transport. When a piece of frog skin is clamped between two chambers containing aerated Ringer's solution, a potential difference across the membrane can be detected by placing the salt bridges of two calomel electrodes close to the membrane on each side. The inside of the frog skin is positive, and the potential difference is maintained for some hours. Knowing that the frog skin enables the animal to take up salt from a low external concentration, Ussing and Zerahn decided to investigate the mechanism of this transport under conditions in which the normal environmental asymmetry was eliminated. They therefore added two further electrodes to the apparatus, Fig. 2.5, together with a potentiometer which could be used to apply across the frog skin a potential difference in opposition to that normally maintained by the skin. By addition of radio tracers to the Ringer solution on one side or other of the skin, the net flux of ions from outside to inside could be studied in the absence of any overall electrochemical potential difference. In order to make this idea clearer, Fig. 2.5 also shows a magnified diagram

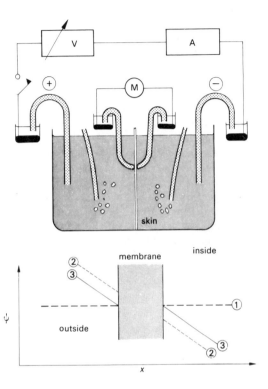

Fig. 2.5 Short circuit apparatus for frog skin. V is variable voltage supply, A is ammeter and M measures membrane potential. Lower diagram shows potential ψ against distance x. Profile in membrane not shown. For Cases 1,2 and 3 see text.

of electrical potential against distance in the neighbourhood of a thick membrane in Ussing's apparatus. Cases 1 and 2 in the diagram concern a non-living, ion-permeable membrane. The biassing potentiometer is switched off in Case 1, therefore there is no potential gradient (or field). When the potentiometer is switched on the membrane is placed in a constant electrical field; though the potential profile within the membrane cannot be predicted, the sign of the potential difference must be as shown. In Case 3 the membrane represents the frog skin in an applied field adjusted so that there is no potential difference across the membrane. Since there is a potential difference in the solution there must be a current, and in fact this is why the biassing circuit includes an ammeter. When the preparation is adjusted as shown in Case 3, the current in the external circuit is called the 'short circuit current'; Ussing's discovery was that when the net flux of Na^+, found from the difference between the unidirectional influx and efflux of radio tracer sodium, was converted into electrical units it was equal within experimental error to the short circuit current. Since there was no potential difference across the skin, this meant that the whole 'current' must be carried by some biochemical mechanism capable of the translocation of Na^+ and powered by a supply of energy within the membrane. In confirmation it was found that the quite large unidirectional Cl^- fluxes were equal within experimental error.

In other experiments net influx of Na^+ was shown between a tenfold diluted Ringer on the outside, and normal Ringer on the inside, under which conditions the inside of the skin was positive with respect to the outside. After addition of dinitrophenol, the flow of sodium was reversed, the skin behaving like a non-biological membrane. Experiments such as this are consistent with a widely accepted definition [32] of active transport as the net flow of a substance against an electrochemical gradient. The biochemist may however by more struck by the fact that in the same system under different conditions, net transport occurs in the absence of any overall gradient. Another point of view of the term active transport is offered in Section 3.1.

The short circuit current technique has been applied with success to a number of tissues for the measurement of ion fluxes. The voltage clamp technique for the investigation under controlled conditions of membrane currents in nerve was developed independently [20] at about the same time as the short circuit technique; it is a brilliant application of the same principle, being much more sophisticated because of the need to cope with very rapid changes of current. More relevant to the present book are the measurements of short circuit currents across pieces of isolated intestinal mucosa, which were undertaken in order to investigate possible relationships between the transport of non-electrolytes (sugars and amino acids) [35] and sodium in intestinal absorption. It was shown that the increase of the short circuit current from the mucosal to the serosal side was a rectangular hyperbolic function of the concentration of an amino acid (e.g. alanine, valine, or leucine) on the mucosal side, suggesting saturatable catalysis of sodium transport by an amino acid. By means of tracers the unidirectional influx of amino acid into the epithelial cells was measured at varying amino acid and sodium concentrations in the mucosal solution; double reciprocal plots of the results were straight lines, that of lowest slope (corresponding to highest flux rate) being at the highest sodium concentration. The apparent K_m for amino acid uptake decreased with increasing sodium concentration, the maximum flux rate being unaffected by the sodium concentration. One possibility raised by these experiments and by similar findings with sugar transport, is that uptake of the non-electrolytes occurs at the same sites as that of sodium. Results obtained by other techniques do not all support this conclusion; however the clarity of the *in vitro* work summarized here makes it very likely that co-transport of sodium and non-electrolytes is an important element in the complex mechanism of intestinal absorption. Evidence for similar types of co-transport has been found in some studies of single cells, notably pigeon erythrocytes.

An important principle is raised by the result of another experiment in the same field. The short circuit current across isolated rabbit ileum was more than doubled by the addition of 3-O-methyl glucose, a non-metabolized sugar: this eliminates the possibility that an autocatalytic mechanism in which the non-electrolyte acts as an energy source underlies the findings of co-transport. It will probably not surprise the reader that phlorizin reduces the short circuit current to the basal level in the absence of the sugar.

Having seen many examples of active transport in absorption processes, we come to a brief consideration of secretion. Amongst the many physiological examples, gastric secretion is particularly dramatic, and there is no difficulty in classifying, as an overall process of active transport, the secretion of roughly 1·7 M hydrochloric acid from approximately neutral raw materials. On the other hand it is virtually impossible to give a coherent summary from available evidence which might suggest a mechanism; quite apart from the cellular complexities involved, the major reason is that the technique of specifically labelling the cation concerned is inapplicable. One possibility to which much attention has been devoted, is that protons derived from metabolic substrates (and therefore in many cases indirectly from water) are released outside the secretory membrane. At the simpler end of the scale, acid is produced in many microbial cultures, for example in the 'mixed acid fermentation' by E. coli. It is interesting that in bacteria many metabolically important dehydrogenases are membrane bound, [30] though this in itself does not implicate them in the secretion of protons. In a number of experiments with unicellular organisms it has been possible to demonstrate unambiguously one for one exchange of H^+ with K^+. The possible significance of exchange of other cations with H^+ is considered further in Section 3.1.

A major advance in the study of active transport was heralded by Skou's discovery of an ATPase in crab nerve which required Mg^{++}, and was activated by sodium and potassium. Similar enzymes were subsequently investigated thoroughly in kidney and red cells, were shown to be specifically inhibited by ouabain, and have been identified with the sites for active transport of sodium. This subject is discussed in more detail in Section 4.3.

2.3 Ionic distributions

This section consists of a table, which compares the ionic composition of rat skeletal muscle with that of the plasma, together with a few comments on the data. It is suggested that the reader gives thought to the problems presented by these data. A possible interpretation is given in Section 3.3.

Table 2.2. Ionic composition of rat skeletal muscle [8]

Species	Fibre water mM	Plasma mM
K	152	6·4
Na	16	150
Ca + Mg	18	5
Cl	5	119
HCO₃	1·2	24·3
Phosphocreatine	37	
ATP	10	
Hexose phosphate	15	
Triose phosphate etc.	5	
Acid soluble phosphate	5·5	2·3*
Anserine and carnosine	33	
Protein	4	1
Amino acids	30	3
Lactate	3	1
Urea	7	7
Glucose	—	5

* Inorganic phosphate.

The figures in the table, which are expressed per volume of fibre water, and derived from the analysis of whole muscle and have been corrected for the presence of ions in the extracellular compartment, the composition of which is not open to experimental determination but is calculated from the composition of plasma and the volume of the extracellular compartment. This results in some uncertainty in the figures for the first group of species, the inorganic ions.

The second group, the phosphate esters plus inorganic phosphate do not constitute pure fractions, being distinguished from one another analytically according to their stabilities. All are negatively charged.

Anserine and carnosine are dipeptides, cationic at neutral pH. It is difficult to analyse for glucose in muscle, but the concentration may not be zero.

Figures for phospholipids have not been included because these compounds do not enter into the osmotic balance of the tissue with its surrounding fluid and thereby with the plasma. Other substances in muscle may have been missed in the analysis. Certainly the inequality of the total figures should not lead us to believe that muscle is in osmotic imbalance with the plasma. In fact on the evidence of the greater depression of freezing point of freshly excized and ground tissue compared with that of plasma, it was suggested that living muscle could only maintain its evident osmotic balance by continually pumping out water. Careful experimental work revealed no evidence for such active transport of water, and it has since been demonstrated, by careful technique, that the freezing point depression of some tissues is equal to that of plasma. Experimental departures from this equality are due to rapid tissue autolysis, which increases the number of molecules present.

The membrane potential of rat skeletal muscle is 90 mV, outside positive (see Section 4.1).

Bibliography
Davson, H. (1970) *A textbook of general physiology* 4th edition. Churchill, London.
The emphasis of this thorough book, the third edition of which is still useful, has always been on transport. A good source of references.

26

3 Concepts underlie models

3.1 Vectorial metabolism

Living cells require fuels which must be able to get in, and they produce end products, including water, which must be able to leave; in growing cells the input exceeds the output, whereas in the metabolic steady state (usually chosen for transport experiments) input and output are equal. Cells need to control their internal concentrations of the small molecules, or anions, which begin metabolic sequences, and in some cases small molecules which are products of metabolic sequences; control of these concentrations is an essential contribution to the regulation of the intervening metabolic sequences. Cells also contain ions such as Na^+, K^+, and Cl^- which, by their nature, cannot enter into metabolic reactions to give changed products, but may partake in catalysis as activators or inhibitors. The specific entry mechanisms which have evolved for these various chemical species may be understood by extension of the principles of metabolic biochemistry, in particular extension of the principle of the common intermediate (coupling principle) so as to understand the consequences of the steps of a cyclic reaction sequence being spatially distributed in an anisotropic medium (i.e. a membrane). A principle not found elsewhere is that a chemical reaction which entails translocation may be coupled with the partial dissipation of an electrochemical gradient; the maintenance of such a gradient, which is necessary in the steady state, must ultimately depend on a primary process of ion secretion.

3.1.1 Cells must catalyse their fuel uptake.
From Collander's values (see Section 1.3) for the permeability of the cell membrane to non-natural substances, one can calculate the permeability expected of a fuel such as glucose if it entered the cell by a similar mechanism to the probe molecules. For example, imagine 1 ml of cells, having an area A of 10^3 cm^2 and a permeability coefficient k for hexose, extrapolated from Stein's graph, of about 6×10^{-5} $cm.h^{-1}$, bathed in 5mM hexose, then the rate of hexose entry would be given by

$$\text{uptake rate} = kA\Delta c = 3 \times 10^{-7}\,mol.h^{-1}$$

provided that the hexose concentration inside was effectively zero as a result of its metabolism. Now compare the hexose utilization required to maintain a reasonable Q_{O_2}, let us say 22·4 $\mu l.mg^{-1}$ (dry) h^{-1} for convenience. Suppose cell dry weight is 20% of wet weight, then since 6 molecules of oxygen are needed per hexose

$$\text{utilization rate} = \frac{Q_{O_2} \times \text{dry wt (mg)}}{22 \cdot 4 \times 10^6 \times 6}$$

$$= 3 \cdot 3 \times 10^{-5}\,mol.h^{-1}$$

Clearly the cells would starve unless the entry of fuel was catalysed. The ubiquity of saturable entry processes makes it hard to believe such catalysis was an evolutionary afterthought: more likely it evolved along with enzymic catalysis as a variant of it.

Mitchell proposed the concept of 'vectorial metabolism' to encompass the many elements

27

of metabolism in which direction is inherent as well as magnitude; uptake processes are prominent amongst these elements. The idea of directional chemical catalysis is unfamiliar from non-biological chemistry, except surface chemistry, though it is inherent in Eyring's theory [11] of absolute reaction rates. However orientation effects and conformation changes are becoming familiar in enzymic catalysis, and the extension of these ideas to cope with passage through a membrane is qualitatively acceptable. While the concept of vectorial metabolism desirably avoids arbitrary mechanistic postulates it has a valuable suggestive generality. As an example Fig. 3.1 compares a postulate [26] (1961) of Mitchell's for the transfer of a group G into a cell by means of a translocase, with the vectorial arrangement (reported in 1966) [22] for the transfer of one of a number of sugars into bacteria by a translocating phosphotransferase system. In the 1961 postulate, the translocase E is an enzyme which participates in the following sequence:

(a) Scission of the bond between the group G and its metabolic donor D.
(b) Translocation of G
(c) Formation of a covalent bond between G and the acceptor A.

One may think of the translocase E_{II} (Enzyme II) in the bacterial system as corresponding with the postulated E. The following comparable steps are catalysed by E_{II}:

(a) Acceptance of the sugar Sx from the medium by E_{II}.
(b) Translocation of Sx.
(c) Formation of and release of phosphorylated sugar Sx P.

Enzyme II is magnesium dependent, and is one of a set of enzymes each specific for a different sugar; however the phosphorylated histidine containing protein (P)HPr is a substrate for all Enzymes II, and may be thought of as analogous to the acceptor A. In completion of the process of entry of the sugar, (P)HPr is dephosphorylated,

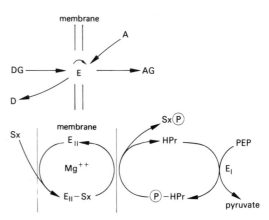

Fig. 3.1 Postulated vectorial enzyme (top) compared with bacterial phosphotransferase system (bottom). See text for details.

and to return the system to its initial state HPr is rephosphorylated by phosphoenol pyruvate (PEP), the reaction being catalysed by a soluble phosphotransferase called Enzyme I (E_I). Notice that in this example the endergonic process of sugar phosphorylation and the potentially endergonic transfer of sugar into the cell at sufficient concentration for the metabolic demand are coupled with the exergonic dephosphorylation of phosphoenol pyruvate. Coupling with an exergonic reaction is inherent in the mechanism of this catalysis, though one cannot tell whether it is necessary in order to achieve a sufficient rate of entry. In this example covalent bonds are broken and made whereas in many translocations, for example those of sugars or amino acids into red cells only non-covalent interactions are involved; in such cases the vectorial catalyst is not described as an enzyme because its 'product' is the unaltered substrate in a different place.

3.1.2 The carrier postulate and metabolic coupling.

In order to explore further the dependence of biological transport on the coupling of reactions, it is useful to separate conceptually the processes of translocation and chemical reaction which underlie metabolically linked transport. Such separation is traditionally postulated in the 'carrier' interpretation of biological transport which, though it predates Mitchell's concept, is still much used. The relationships described in Section 2.3 for the transport of sugars in red cells have been explained by postulating mobile carriers which can circulate in the membrane. As indicated in Fig. 3.2 the carrier site can cross the membrane phase in two forms, loaded or unloaded. This form of transport has been called 'facilitated diffusion'. In order to avoid encouraging the view (which may well never be proven) of carriers like submarines ferrying passengers through a layer of oil, it is well to concentrate on the mechanistic essentials of the postulate for which there is widespread experimental support; there are clearly saturable carrier sites, analogous to the active sites of enzymes, which have alternative access to one or other side of the membrane, and may have different affinities for the same substrate when it is presented from contrary sides (see Section 4.3). In ignorance of the molecular details it is reasonably assumed that the carrier site is part of a macromolecular system which catalyses translocation; Mitchell has usefully described the whole system, of which the carrier site is a part, as a *porter*.

The most important quantitative properties of this model are summarized in terms of its parameters which are symbolized in Fig. 3.2. The two sides of the system are indicated by ′ and ″. The reactions of the site C are described by conventional equilibrium constants K_1 and K_3 (dissociation constants) and K_2 and K_4 (distribution constants). Like all chemical reactions each step is in principle reversible, and similarly the

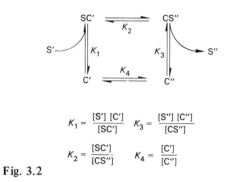

$$K_1 = \frac{[S'] \, [C']}{[SC']} \qquad K_3 = \frac{[S''] \, [C'']}{[CS'']}$$

$$K_2 = \frac{[SC']}{[CS'']} \qquad K_4 = \frac{[C']}{[C'']}$$

Fig. 3.2

membrane in which the porter is set is in principle permeable to the substrate S. After letting the porter system come to equilibrium with its substrate, the system can be tested thermodynamically by transferring one mole of S from the left to the right through the porter, and returning S uncatalysed, through the membrane instead of the porter; finally the unloaded carrier is returned to the left. The system has now been restored exactly to its initial state, so the total change of free energy in the cycle must be zero. Since the uncatalysed path can only be in equilibrium when $[S'] = [S'']$ it follows that the accumulation ratio for the porter in equilibrium is

$$\frac{K_3 K_4}{K_1 K_2} = \frac{[S'']}{[S']} = 1$$

Another way of putting this is to say that since a catalyst cannot alter the position of equilibrium, the overall equilibrium constant for the porter system must be unity. This relation shows that three of the four equilibrium constants are independent, as are seven of the eight rate constants which may be defined. It follows that the ratio of the dissociation constants is equal to the ratio of the distribution constants:

$$K_1/K_3 = K_2/K_4$$

This means, for example, that if $K_4 \to \infty$ (i.e. the unloaded carrier is thermodynamically stuck on

29

the outside) $K_2 \to \infty$ also, so that one of the ground rules for the carrier game is that both the loaded and the unloaded forms must always be shown.

Now let us suppose that translocation of S through the porter is followed by metabolic transformation of S'. The previous test of the porter showed that its standard free energy $\Delta G^0_{\text{translocation}} = -RT \ln K = 0$. Therefore for the overall process, of translocation followed by metabolism, to be spontaneous, $\Delta G^0_{\text{reaction}}$ must be negative since $\Delta G^0_{\text{translocation}} + \Delta G^0_{\text{reaction}} < 0$ for spontaneity. Under the biochemist's ground rules for the thermodynamics game, this statement implies that in the steady state, the coupled reaction pulls the translocation in favour of its product, a kinetic extrapolation which is only acceptable because biochemists believe that evolution has provided the catalysts which permit certain thermodynamically spontaneous reactions to go at biologically desirable rates (see Section 5.1).

The simple scheme of Fig. 3.2 accounts for the findings in a number of experimental systems, such as those described in Section 2.1. In the course of attempts to explain other experimental results on transport which are not consistent with this postulate, many other hypothetical coupled porter schemes have been explored. Such attempts have often suffered from the tendency to explain the results in terms of schemes in which the number of parameters exceeds the number of independent experimental measurements for their determination, whereas what should be established is the minimal scheme [17] which will explain the results of different experimental approaches.

Now imagine a situation in which a sequentially coupled porter cannot deliver its substrate fast enough for the cell's needs, and suppose that this defect can be remedied by modification of the porter hypothesis, the carrier site itself becoming a metabolic intermediate, so that the porter is directly, instead of sequentially, coupled

30

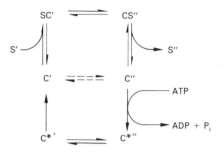

Fig. 3.3

with metabolism. A conventional postulate of this kind is shown in Fig. 3.3. A new form C* of the carrier site appears as a result of a reaction in which ATP is split; there are now two alternative cyclic sequences for the translocation of S, because there are two different ways in which the unloaded carrier can return to its starting point. For simplicity, let us first consider the extreme case in which all of the unloaded carrier returns as C* (that is, dotted arrows indicate zero rate). The net result of this 'fully coupled' cycle is the hydrolysis of a mole of ATP and the translocation of a mole of S. This can be summarized

$$S' = S'' \qquad \Delta G_{\text{translocation}}$$
$$\underline{ATP + H_2O = ADP + P_i \qquad \Delta G_{\text{reaction}}}$$
$$S' + ATP + H_2O = S'' + ADP + P_i \quad \Delta G_{\text{total}}$$

because all other parts of the cyclic reaction scheme cancel out in the addition. If there were no other route for G through the membrane, the system would reach equilibrium in which the rates through the backward and forward paths would be equal. At equilibrium

$$\Delta G_{\text{total}} = 0$$

$$\therefore \Delta G \text{ translocation} = RT \ln \frac{[S'']}{[S']} = -\Delta G_{\text{reaction}}$$

Since $\Delta G^0_{reaction}$ is about 30 kJ. mol^{-1}, the accumulation ratio would be large (the reader may calculate the value).

To be more realistic, the membrane would itself be permeable to S, and some of the unloaded carrier would return in the form C. For this system to have a biological advantage over the sequentially coupled porter, it might be well to postulate that the movement of C was much slower than that of CS or C*; the mole ratio of transport to ATP splitting would then be only just greater than 1. In a cell the coupled porter would not reach equilibrium but would run cyclically to accumulate S at a rate set by kinetic considerations, principally the rates of resynthesis and supply of ATP.

Depending on the conditions there is a wide range of possibilities for the operation of this kind of cycle, which may effect 'active transport'. Two examples will suffice: if the substrate is not metabolised it will accumulate to an internal concentration at which its total efflux, by the reverse path through the porter and any slight leakage through the membrane phase, equals its influx by the forward translocation reaction; in contrast if it is metabolized at a rate equal to that of the forward translocation reaction, and if the first enzyme has the same K_m and V_{max} as the outside of the porter, the inside and outside concentrations of the substrate will be equal. No useful purpose is served by defining here the equilibrium constants in Fig. 3.3, since all that one can conclude thermodynamically is summed up by the statement that for the reaction cycle to bring about translocation

$$\Delta G_{total} = \Delta G_{reaction} + \Delta G_{translocation} < 0$$

And, more generally, no useful purpose is served by protracting the search for an operational definition of active transport which would encompass the possibilities of the directly coupled porter: it is better to look for the evidence which would identify the mechanism, name it clearly, and abandon the term 'active transport'.

3.1.3 Variations on the cyclic theme

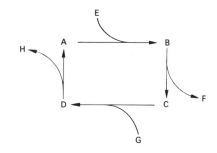

Fig. 3.4

A large part of metabolic biochemistry can be analysed in terms of cyclic pathways. Fig. 3.4 shows a generalized reaction cycle catalysed by at least 2 enzymes. Amongst its normal functions is the performance of the spontaneous reaction

$$E + G = F + H$$

with the catalytic participation of the free intermediates (A,B,C, and D, if there were four enzymes). The free energies of the individual reactions can be positive or negative provided their sum round the cycle is negative; thus endergonic biosynthetic reactions are often coupled cyclically with exergonic degradative reactions. But if the sole function of the cycle is transport (as in the previous postulate) degradation is coupled with the production of the same substance in a different place, so the scheme should really be compared with a purely degradative cycle, if such exists. The closest familiar example is the futile cycle, Fig. 3.5, which arises from the inevitable juxtaposition in many cells of conjugate degradative and biosynthetic paths. Its net result is the hydrolysis of ATP and (since the two enzymes cannot simultaneously be in the same place) the translocation within the cell of the phosphate group.

In this example ATP is hydrolysed in two distinct steps. Chemical commonsense and the

31

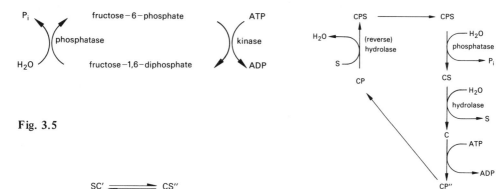

Fig. 3.5

Fig. 3.7

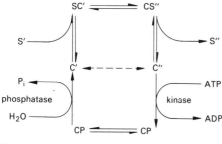

Fig. 3.6

vain search of the known pathways of intermediary metabolism for a case like that of Fig. 3.3 in which the hydrolysis occurs in a single step, justify a modification of the carrier scheme (Fig. 3.6). A major difference between Fig. 3.5 and the bottom half of Fig. 3.6 (reaction and transport) is that in the latter, a special substrate, the carrier site, is localized in the membrane phase instead of free in solution, and the total amount of its different forms is constant. One could even change one's interpretation of Fig. 3.6, and think of the bottom half as catalysed by a phosphatase and a kinase in apposition across the membrane; one must then think about the top half in terms of vectorial metabolism.

A biological disadvantage of this chemically realistic scheme becomes apparent; for every S taken in, a P_i is lost. This could be rectified by

32

introducing (Fig. 3.7) a third enzyme: or, if S binds to C non-covalently, by omitting the two hydrolases and considering S' as the external activator of a vectorial phosphatase. This viewpoint is consistent with the absence of any evidence to date of covalent linkage between a carrier site and a substance (sugar or amino acid) which is delivered unmodified on the inside. It is also relevant that the Na/K antiporter ATPase (see Section 4.3) is associated with phosphatase activity, the natural substrate of which is not identified.

In summary, these examples are intended to reinforce the suggestion that reaction and transport need not in reality be separated, and to stimulate further consideration of vectorial metabolism by looking for analogies between transport and intermediary metabolism.

3.1.4 Kinetic aspects of transport.
Much of our understanding of intermediary and vectorial metabolism comes from kinetic studies, but we know much less about transport mechanisms than enzyme mechanisms partly because of the greater technical and conceptual difficulties of kinetic studies of transport. The purpose of this section is not to grapple with these difficulties, which are fundamental in contemporary

research, but to offer brief comments intended to ease the reader's approach to the literature which has already accumulated.

The idea of the rate determining step is important in kinetics. For example in a linear reaction sequence of

$$\rightarrow A \rightarrow B \rightarrow C \rightarrow D \rightarrow$$

unidirectional steps in a steady state, the rate of 'throughput', is determined by a single step only if its rate constant is about 1/10 or less [6] than that of any other in the sequence. In complicated reaction schemes the rate determining step (or group of steps) often depends on the experimental conditions.

If a variety of kinetic experiments are done under different conditions on biologically identical systems, (having carrier sites on the same nature and concentration), then a satisfactory model for the transport system is one with which all the experiments are consistent, and all the parameters of which are unequivocally determinable from the experiments. Some confusion seems to have arisen because this desirable state of affairs has been hard to achieve in practice. For example, it has been a common observation that after the addition of a trace of labelled substrate to cells plus unlabelled substrate, which have reached a steady state, 1:1 exchange of labelled and unlabelled substrate occurs without net transport. This has been described as transport by *exchange diffusion*, and has been attributed to the operation of the carrier scheme shown in Fig. 3.8. However, the result could be explained on the model of Fig. 3.2 (at equilibrium) or that of Fig. 3.3 operating concentratively in the steady state in parallel with leakage of the non-metabolite S through the membrane; indeed if the labelled substance were water or one of certain non-natural permeants there would be no need to postulate a carrier. It is safest to think of Fig. 3.8 as the maximum contribution to a transport model which is justified by the demonstration of the *phenomenon* of exchange diffusion, taken by itself.

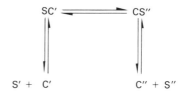

Fig. 3.8

If cells are equilibrated with a labelled non-metabolized substrate at a concentration which nearly saturates the carrier site, and when an accumulation ratio of 1 is reached, unlabelled substrate, or a related substrate, is added to the medium, efflux of the labelled substrate against its concentration difference is observed for a while. The reason for this *counterflow* is that, for a while, competition for the carrier site by the external substrates is not balanced by competition on the inside, therefore labelled substrate is carried out faster than it comes in. Such observations of counterflow also suggest that, if two substrates share a common carrier, a maintained gradient of one may enforce translocation of the other. This idea will be further explored in a different context (Section 3.1.5).

In the scheme of Fig. 3.2 it can be shown that with substrate concentrations below the saturation level, the unidirectional influx rate of a substrate may depend on the concentration of S'', for example a decrease of $[S'']$, which increases the trans-membrane concentration difference for S, paradoxically results in decreased influx. This *trans-concentration effect* may be explained as follows. The influx rates depends on the availability of carrier sites on the outside, which depends on the rate of return of the carrier to the outside. If the equilibration of substrate (steps 1 and 3) is fast compared with the reorientation of carrier (steps 2 and 4), then decrease of $[S'']$ will raise the concentration $[C'']$, and this should increase the return rate of the form C. But if in fact the rate

33

constant for CS return is greater than that for C, reduction of $[S'']$ will, by reducing $[CS'']$ also reduce the return of the carrier by this faster route (to be unloaded outside). If this factor outweighs the increased return of the C form, the trans-concentration effect is observed. For these reasons the phenomenon has been called *accelerative exchange diffusion.* This argument is strong because it has been shown unequivocally [18] that for the transport of a non-electrolyte into red cells, reorientation of carriers is rate determining. Indirect evidence that this state of affairs is widespread comes from studies of the biological selectivity between ions (Section 3.4).

3.1.5 Transport coupled to primary ionic secretion.

It has been usual to postulate ATP hydrolysis as the exergonic reaction coupled with translocation, but the only unambiguous demonstration of ATP as the direct energy donor for transport, has been for the Na/K antiporter ATPase system. On the contrary, it has become apparent that ATP is not directly involved in a number of metabolically linked systems; instead the driving force is provided by a difference of electrochemical potential maintained by metabolism. A porter couples spontaneous cation transport in the direction of decreasing electrochemical potential with the translocation of the substrate, which may be neutral or electrically charged. For example in the uptake of β-galactoside by *E.coli* (Fig. 3.9) [42] co-transport (symport) of a proton and the substrate occurs, and the return of the unloaded carrier in the electropositive direction is assisted by its negative charge with respect to the loaded form.

The concentration difference of the driving cation is maintained by another process, primary ionic secretion, the free energy for which must ultimately derive from the free energy of oxidation of reduced compounds. The most direct way of providing this link would be to

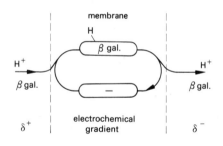

Fig. 3.9

suppose that the protons produced by the metabolism of substrates are released at the outside of the membrane within which metabolism occurs. This idea, of which Conway was an early proponent, was generalized by Davies and Ogston [9] and extended by Mitchell. The idea that the proton is more suitable for the primary transport process than a cation such as Na^+ or K^+, which nearly always occur in soluble salts, makes chemical sense and is physiologically rational because the buffering capacity, under any likely environmental conditions, ensures the cell of an almost inexhaustible supply of the primarily transported species without the build up of high concentrations. Alternatives to H^+, as a metabolically produced primary secretion, are OH^-, NH_4^+ and HCO_3^-. In a number of mammalian systems, for example the intestinal mucosa, symport of sodium with glucose or an amino acid, appears to occur; the sodium gradient is maintained directly at the expense of ATP hydrolysis, and functionally Na^+ must be thought of as the ion primarily secreted even if the mechanism should prove to involve protons.

In order to maintain electrical neutrality the primary cation secretion must result in the secondary transport of other charged species: extreme cases are that each cation secreted is accompanied by an anion or is exchanged for another cation. In the cell, the charged species which are secondarily transported are the charged sites of porters and probably hydrated cations and anions passing through aqueous channels. The porters may catalyse symport (as in Fig. 3.9) or antiport (e.g. K^+/H^+). This general type of coupling is presumably economical because of its versatility.

3.2 Cells must regulate their osmotic pressure

characteristics while growing and dividing, the metabolic reactions must go on in a co-ordinated fashion. The boundary must therefore be a membrane able to select molecules needed, and eliminate those for which there is no further use.

Natural phenomena, winds blowing, rain falling, and so on, moved cells from one environment to another; sometimes the chemical com-

Table 3.1. Ionic concentrations in sea water

Ion	Cl^-	Na^+	$SO_4^=$	Mg^{++}	Ca^{++}	K^+	CO_2*	Fe	P_i
mEq/kg	535	454	28	52	10	10	2·1	0·004	0·002

* total, mostly as bicarbonate.

In the beginning, life arose in water. The proportions and concentrations of ions in primeval water are presumably similar to those now found in sea water.

Table 3.2 shows the proportions of some elements in the earth's crust generally. Taking the tables together they indicate that the minerals of the earth's crust differ widely in solubility.

Table 3.2. Relative abundance in the earth's crust

Element	Na	Ca	Fe	Mg	
g atoms/kg	1·22	0·90	0·90	0·86	
Element	K	C	P	S	Cl
g atoms/kg	0·67	0·025	0·04	0·016	0·006

The earliest living form which we could fruitfully discuss is a primitive cell, an entity capable of growth and reproduction. One thinks of a primitive cell as a region enclosed within and defined by a boundary, having a small number of very large molecules, DNA or something like it as a genetic material, and working templates of single-stranded RNA; and something similar to what we know as ribosomes. The cell would contain a much larger number of the smaller molecules involved in metabolism and biosynthesis. For the cell to retain its

position of these regions differed, particularly in the amount of dissolved matter, which we could conveniently characterize as the osmolarity. We can rationalize that selection favoured organisms which could cope with environmental changes. If the balance of metabolic and biosynthetic reactions were to be maintained it would prove inconvenient if the selective properties of cell membranes were such that the cells behaved as perfect osmometers. We can see this by means of an example. Suppose that when a cell moved from a dilute to a concentrated medium its volume was halved. The rates of reactions between small molecules are generally proportional to the product of their concentrations, so there would be up to a four-fold increase of these reaction rates. For very large molecules it is realistic to think in terms of the amount of reactive surface available in the cell rather than the molecular concentration, and this surface would be unaltered by change of cell volume. The rates of reactions between small and large molecules would therefore be increased roughly two-fold. A balance between the two kinds of reactions would be very hard to achieve. So one might suppose that successful cells would be those in which mechanisms arose allowing maintenance of constancy of cell volume despite changes of medium osmolarity.

35

One method of arranging this would be to surround the selective membrane with a non-selective open mesh cage of sufficient strength, and to have means of regulating the internal osmotic pressure so that it would always be a reasonable amount more than that outside. The selective barrier would then be held by a hydrostatic pressure against the surrounding open mesh work. If the external excess pressure exceeded the strength of the mesh there would be a catastrophe; but if the regulatory mechanism allowed no internal excess the system would have to respond instantly to any slight internal osmolarity decrease, otherwise there would be a shrinkage. The maintainance of a reasonable difference would therefore give time for a response to all but the most artificially sudden change of environmental osmolarity.

What kind of substance would be most convenient for a cell to use in order to maintain an internal excess osmotic pressure? Given that for 'economic' reasons it should have a low molecular weight, there seem to be two possibilities. The cell could make an osmotic agent according to need, or on the other hand expel it or polymerize it to decrease the internal osmolarity; but this agent would have to be distinct from the nutrients used, or metabolites present internally. Alternatively the cell could pump in, or out, readily available inert constituents common to any environment in which it might find itself.

The first possibility implies regulatory autonomy, but seems energetically wasteful, and the protected metabolic status of the osmotic agent would present a serious difficulty. A further disadvantage of this first proposal would be that the osmotic agent would have to be neutral, because the intracellular creation of a cation, to balance the negatively charged anions necessary for metabolism, would produce an equivalent amount of anions. The second possibility is attractive, not only because it makes use of ubiquitous minerals, but also because the

36

regulatory mechanism could conveniently be made part of the selective barrier, thus allowing simultaneous access and sensitivity to the environment, and the region whose composition is to be regulated.

In either case, the concentration of the osmotic agent in the cell would need to be regulated in a manner sensitive to minute changes in cell volume so as to counteract these changes rapidly. How could this be done? Biochemical sensitivity to a purely physical correlate of cell volume is hard to imagine, but an acceptable alternative is sensitivity to the concentration of a substance, or substances, which determine the rates of reactions in the cell. We must however bear in mind that if the concentration of these regulatory substances were to suffer changes resulting from a cause such as the altered availability of nutrients in the environment, there could be complications calling for some secondary control mechanisms. But if metabolism is versatile, it should be possible to choose intermediates common to a variety of metabolic pathways. Substances such as ATP would seem to be likely candidates.

All this is speculative because nothing is known about primitive cells; the best we can do is consider the facts about present day microorganisms. Although there are many gaps in the evidence, it appears that most microorganisms have no detectable requirment for sodium, and there is evidence that bacteria maintain approximately constant volume by the inward active transport of potassium from the medium. We may perhaps accept, as a working hypothesis, the idea that the first osmotic agent chosen by evolutionary processes was K^+, the common alkali cation with the less soluble salts. (In saying this there is no need to abandon the idea that the primary process is likely to have been proton secretion to which the influx of K^+ was coupled by an antiporter). As we shall see in Section 3.3, many animals differ in that their cells pump out Na^+ in order to maintain osmotic balance.

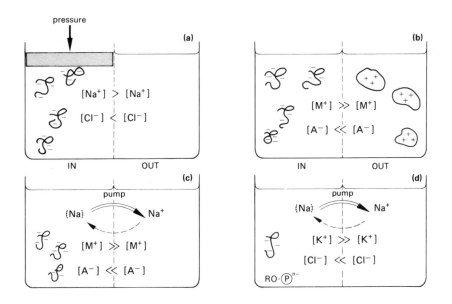

Fig. 3.10 Variations on the Donnan principle. The dotted arrow indicates the leak path. RO—$\textcircled{P}^{n-}$ signifies phosphate ester. See text for details.

3.3 The Donnan principle and the sodium pump

The fluid environment of the cells of animals with kidneys is controlled by the combined activities of different cells and tissues. One result is that there is generally no osmotic difference between the cells and their surroundings, and therefore no need for the cells to be encased in strong meshworks resistant to pressure gradients. This section is about the way in which the postulate of a sodium pump accounts for constancy of cell volume, that is for osmotic regulation, and how it helps one to understand the ionic distribution summarized in Section 2.3. This account, in the tradition of Conway, refers explicitly to the principles of Donnan; their more detailed application is given in Sections 5.2 and 5.3.

A Donnan equilibrium is set up when, for example, a solution containing a negatively charged protein retained by a semipermeable membrane is allowed to equilibrate with a salt solution the components of which can pass through the membrane (Fig. 3.10a). For equilibrium it is necessary to apply to the protein solution that hydrostatic pressure which arrests solvent flow through the membrane; the osmotic pressure of the system is less than that of the protein solution dialysed against water. The distribution of the mobile anions and cations is unequal, for example if they are Na^+ and Cl^-, $[Na_i]/[Na_o] = [Cl_o]/[Cl_i]$ where the subscripts refer to inside (containing protein) and outside. The quantitative interpretation rests on two important principles:

(i) That strict electroneutrality is maintained [15]. Potential differences of less than 1 volt do occur, but the non-uniformity of ion concentration from which they arise is analytically undetectable.

(ii) That each permeant component is so distributed that its chemical potential is the same on both sides of the membrane. (For mobile

37

ions, the term component refers to ions taken together as if they constituted salts).

The Donnan equilibrium is itself of very limited application in biology. It can be used to calculate the composition of the glomerular filtrate (or any ultrafiltrate of plasma) knowing the composition of plasma. It also accounts for the relationship of $[K^+]$ and $[Cl^-]$ between cells and medium in stationary phase bacterial culture. However the two underlying principles are directly applicable to dynamic situations, for example the chloride (or Hamburger) shift in red cells, and, as will now be shown, the effects of the sodium pump.

As a preliminary model. consider a double Donnan equilibrium (Fig. 3.10b) in which polyanion and polycation solutions are separated by a membrane to which they are impermeant. The permeant ions M^+ and A^- are more unequally distributed than they would be if one type of polyion were absent. Concentrations of the polyions can be found at which no hydrostatic pressure difference need by applied for equilibrium. To recapitulate the basic principle in another form: the unequal distribution of the mobile ions results from the asymmetrical distribution of the 'fixed' charge born by the polyions, an asymmetry which in this case is caused by the impermeability of the membrane to these ions. But asymmetry of charge giving rise to a Donnan effect could occur without a membrane, for example slices of two gels soaked in salt solution could be apposed, the cross-linked polymeric material of one gel being anionic and the other cationic. In these two examples the fundamental asymmetry is maintained in different ways, but the resulting distribution of mobile ions may be identical. It would equally be possible to create a cationic asymmetry, and hence a 'double Donnan distribution' (not an equilibrium but a steady state) as in Fig. 3.10c, by omitting the polycation and inserting in the membrane a mechanism which pumped Na^+ ions into the right hand compartment where they

would attain a concentration at which the rate of leakage back to the left equalled the rate of pumping. As a result of this metabolically maintained asymmetry, the permeant ions M^+ and A^- would be distributed according to a Donnan ratio. To make this into an outline model for the cell (Fig. 3.10d) we will have to recognise that the polyanions in the cell are far from sufficient to account for the functional 'fixed negative charge' which must consist mostly of low molecular weight phosphate esters retained in the cell by vectorial metabolism. To test the acceptability of this model, ratios are calculated from the data for muscle in Table 2.2. Donnan distributions are indicated by the ratios: $[Cl_0]/[Cl_i] = 23\cdot8$, $[K_i]/[K_0] = 23\cdot8$ and $[HCO_{3_0}^-]/[HCO_{3_i}^-] = 20\cdot3$; it is reasonable to argue that the bicarbonate ratio might be slightly reduced by the continuous production of CO_2 in the cell which would prevent bicarbonate reaching its equilibrium distribution. The ratios (expressed in the same way) for two other ions indicate larger effects of metabolic production on ion distribution; the lactate ratio is $0\cdot33$ suggesting that relative impermeability of the membrane for this large ion requires an appreciable concentration difference to drive its efflux, and the H^+ ratio (from glass electrode and other studies) of about 5 may indicate the metabolic extrusion of protons. Taking the two amino acids together they must on the average bear a slight negative charge (as do the muscle proteins), so their lumped ratio of about $0\cdot1$ proves that they are not Donnan distributed as a result of the operation of the sodium pump; presumably their distribution depends on vectorial metabolism. If the divalent cations were Donnan distributed their ratio should be the square of the K^+ ratio; consequently the figures in the table come as a salutory reminder of the over simplification of the present model, which neglects the sequestering of calcium in the sarcoplasmic reticulum and the biologically essential chelation of magnesium by ATP and other phosphate esters. In short, we may think of the large proportion of bound intracellular Ca^{++} and Mg^{++} as decreasing the 'fixed negative charge' constituted by phosphate esters and quite probably other impermeant polyanions which are not included in the table. Since there is no evidence that the cationic dipeptides anserine and carnosine are bound, we can consider them as contributing to the intracellular osmotic pressure but decreasing the 'fixed negative charge'.

Quantitative predictions can be made from the model (Section 5.3). If m is the extracellular sodium concentration, p the concentration of 'fixed' intracellular ions, and n their mean net negative charge, it can be shown that there will be zero osmotic pressure difference ($\Delta\pi$) between the cell and tissue fluid if $p = np = m$. (This is the simplest condition for $\Delta\pi = 0$, but is not unique). Detailed, if slightly speculative, calculations indicate that the data are consistent with $m \approx p$ and $n \approx 1$ (perhaps slightly greater) which is a reasonable validation of the model.

The responses of the model and the biological system to altered conditions may be compared as a further test. Boyle and Conway, [3] in a classic paper, considered the behaviour of muscle in Donnan terms before the postulate of the sodium pump became current. When they added extra KCl to the Ringer solution bathing the muscle, KCl moved into the muscle in order to restore the ionic concentration ratios to reciprocal equality and, as a consequence of this solute movement, water entered swelling the muscle. The quantitative agreement between theory and experiment was very satisfactory. Metabolic inhibitors and cooling would be expected to decrease the activity of the sodium pump and thereby cause an osmotic imbalance leading to movement of water into the cells; Leaf [23] showed that on the 'double Donnan' hypothesis, movement of essentially isotonic NaCl into the cells would be expected to occur, with which the experimental findings concurred.

An outstanding biophysical contribution was the measurement by Hodgkin and Horowicz [21] of the rapid response of the membrane potential of single muscle fibres to the extracellular addition of K^+, before the $[K_i]$ had changed appreciably: at $[K_0] > 10\,\text{mM}$ the experimental results were in excellent agreement with the electrochemical prediction of the Nernst equation:

$$E_m = \frac{RT}{F} \ln [K_i]/[K_0]$$

Underlying this correlation is the implication that the fibre membrane is much more permeable to K^+ than Na^+, which is biologically reasonable if the sodium pump is to be able to maintain such a large asymmetry with a reasonable energy expenditure. The origin and significance of bioelectricity is discussed in Section 4.1.

The prediction that $\Delta\pi = 0$ when $p = np = m$ suggests that osmotic balance could be maintained under a variety of metabolic steady states (and rates) provided that the ratio of the sodium pump rate and the phosphate ester pool remained constant. We may ask whether these two systems both respond to the same metabolic parameter, say [ATP]/[ADP], and perhaps share some regulatory mechanism? The isolation of the Na/K activated ATPase, its identification with the sodium pump, and its sensitivity to the ratio [ATP]/[ADP] are evidence for such a link. It is also necessary to amend the model by stating that there is a mechanistic obligation (antiport) for pumping of some of the Na^+ to be coupled chemically with K^+ influx, in addition to the general electrochemical obligation imposed by the Donnan conditions.

In conclusion, a semiquantitative treatment of the thermodynamic consequences of the postulate of the sodium pump explains the major features of the ionic distribution between cells and tissue fluid in vertebrates.

3.4 Cation selectivity

The unhydrated alkali cations increase in size with increasing atomic number, whereas it is known that the size of the hydrated ions increases in the reverse order (though the hydrated radii are not precisely known). This can be explained on very simple physical principles, that the electric field at the surface of a sphere is proportional to the charge per unit area, and that there is a minimum energy of interaction between the electric field and the water dipoles for hydration to occur; it follows that the greater the field at the surface of the ion, the more molecules of water will adhere. Although this has long been known, many biologists, and even chemists, have felt that such simple principles could hardly account for the variety of the chemical and biological specificities of the alkali cations. It is no longer necessary to suffer from this feeling because the question of the selection between the alkali cations, both in physicochemical and biological systems has been thoroughly studied by Eisenman and his colleagues. With a view to facilitating biological and other measurements, they first devized glass electrodes which were insensitive to change of pH over all but the lowest part of the normal range, but responded instead to the alkali cations, one of them usually predominating over the rest. Two limiting selectivity orders may be recognized (Table 3.3), I, that in which the non-hydrated radii decrease, and XI, the converse order, that of decreasing hydrated radius. Although there are logically 120 possible selectivity sequences including these two, only 11 were observed in practice, and these are shown in the Table 3.3. This was not merely a limitation of the craft of making glass electrodes because the same 11 patterns were found in ion exchange resins, and when Eisenman turned his attention to the biological literature he found that (with one minor complication) the same sequences were apparent in the numerous diverse biological phenomena for which sufficient

Table 3.3. Cation selectivity orders.

I	Cs > Rb > K > Na > Li
II	Rb > Cs > K > Na > Li
III	Rb > K > Cs > Na > Li
IV	K > Rb > Cs > Na > Li
V	K > Rb > Na > Cs > Li
VI	K > Na > Rb > Cs > Li
VII	Na > K > Rb > Cs > Li
VIII	Na > K > Rb > Li > Cs
IX	Na > K > Li > Rb > Cs
X	Na > Li > K > Rb > Cs
XI	Li > Na > K > Rb > Cs

Table 3.4. Cation selectivity in biology.

III	Live *Chlorella* ion exchange
IV	Frog sartorius muscle flux
VI	Blowfly salt receptor stimulation
IX	Squid action potential
X	Frog skin permeability from outside
XI	Cation binding of DNA and RNA

Roman numeral gives order: see Table 3.3

information was available. Table 3.4 lists a few of these. Eisenman's quantitative explanation is here summarized qualitatively. He plotted (actually in terms of free energy) the relative preference of the ion for the solution rather than the ion exchanger, against the 'radius' of the anionic sites. Doing this schematically for three ions (Fig. 3.11), it can be seen that the curves rise at different rates as the field decreases and therefore they cross one another at different points. Each crossing point marks a transition between one selectivity order and the next; as anionic field strength increases the selectivity orders rise from I to XI. The graph expresses the electrostatic competition of water (a polar molecule) and the anionic site for ions of different size. The electrostatic attraction of the ion for the site is proportional to the inverse square of the distance of separation of their centres, but that for the ion and water falls off more rapidly with increasing separation, being proportional to something between the inverse

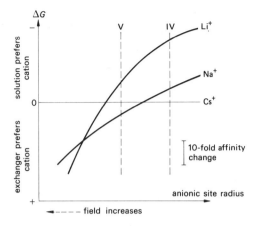

Fig. 3.11 Schematic diagram indicating preference of Li$^+$ and Na$^+$ for solution relative to Cs$^+$. Selectivity order (see Table 3.3) differs at two field strengths shown by dotted lines.

cube and fourth power of the distance. Consequently, if two ions which change places at the transition between the selectivity orders are compared as the field strength is increased, the larger ion will be the first to lose water in favour of the site because its hydration shell is less tightly held than that of the smaller ion.

What are the implications of this theory? The field strength of an exchanger can be increased by spacing the sites more closely or by decreasing the radius at fixed separation. Increasing anionic field strength is correlated with increasing pK, that is the weaker acid sites have the higher field strengths. It is even possible to relate the pK of a site to the selectivity order, subject to an assumption about the spacing of the sites, since it is well known that the electric environment of an acidic group affects its pK. An example of the achievement of the theory is the investigation by Diamond and Wright of the pK of the groups lining aqueous channels in rabbit gall bladder. In one kind of experiment their pK was found to be about 3, which is in the correct range for carboxyls. The selectivity order predicted on the assumption of 'infinitely' spaced sites agreed with that determined from quite different experiments on the tissue.

The outlook for this theory is clouded only by its inapplicability to protons, and by the general lack of any means of determining proton selectivity experimentally.

The fact that the biological selectivities are well correlated with those of non-living systems, for which the underlying principles are quantitatively verifiable, has an important consequence. It suggests that the same types of processes, that is equilibrations of ions with sites, are being compared in the very varied complex biological phenomena, and that the steps which differ between the biological phenomena, for example the rate of translocation of loaded carrier sites, are relatively insensitive to the cation selected. Also, if those steps in which the preference for the different cations is shown were to be biologically rate-determining, a quite different pattern would have emerged.

Bibliography
Mitchell, P. (1967) in Florkin, M. and Stotz, E.H. *Comprehensive Biochemistry.* Vol. 22, Chapter IV. Elsevier, Amsterdam.
A dialectical review of vectorial metabolism, full of useful information.
Smith, H.W. (1953, reprint 1961) *From fish to philosopher.* Doubleday, New York.
Metabolism of salt and water affects everything from prehistoric fishes to the virtuoso pianist's fingers.
Edsall, J.T. and Wyman, J. (1958) *Biophysical Chemistry,* Academic Press, New York.
Chapter 1 is on the geochemical background to biology.
Ruch, T.C. and Patton, L.D. (1965) *Physiology and Biophysics.* Saunders, Philadelphia.
Chapters 1,2 and 43, about ion transport generally, and in nerve and kidney, are particularly relevant.
Eisenman, G. (1965) in *23rd International Congress of Physiological Sciences,* page 489,

Excerpta Medica Foundation, Amsterdam.
A review of the biological literature on cation selectivity.

4 Experimental approaches to mechanism

4.1 The significance of bioelectricity

So far the bioelectric potential differences, commonly detectable across membranes and relevant to the study of ion fluxes, have hardly been mentioned in this book. One reason is that many students get confused by a kind of 'chicken and egg' argument about whether the potential difference* is set up and results in the ion flux, or alternatively whether a biochemically driven ion flux results in a p.d. Since a rational explanation of the basis of ion fluxes can be given without reference to this question it has so far been avoided, and the reader might be tempted to conclude that the second alternative is nearer the truth. But this conclusion is not really justified, and may arise from a misunderstanding about electrochemistry: any separation of electrical and chemical aspects of this subject is arbitrary. Electrochemical processes are those 'chemical' changes in which a transfer of charge occurs and both electrical and chemical aspects must be considered. In fact this is what we are doing when we use the equation $\Delta G = -nFE$ for an electrochemical cell (see Section 5.1). An equation summarizes quantitatively two ways of looking at the same thing; here the left side deals with the chemical aspect and the right side with the electrical aspect. In biochemical thermodynamics both approaches have been used according to their experimental and didactic convenience. So far we have concentrated on the free energy approach, but in view of the heavy dependence of

single cell physiology on potentiometric measurements, we ought to see whether the electrical viewpoint gives us a further insight into biological transport.

Potential differences may be detected across biological membranes by the use of suitable electrodes. The probe electrode is almost invariably a microelectrode, a fine sharp ended glass capillary which can be inserted into cells; the microelectrode is filled with concentrated KCl solution and is really a salt bridge connecting the inside of the cell with a calomel electrode. The reference electrode is generally larger and usually a calomel or silver/silver chloride electrode. The electrochemical cell consists of the two electrodes and their mutual electrical connection through the preparation studied. The reference electrode is placed in a position in the tissue or bathing medium which is unaffected by the experimental procedure. This is important because all that can be measured is a single p.d., and there is no way of determining the contribution to this of any p.d. between the reference electrode and the aqueous medium in which it is immersed. In practice, knowledge of electrochemistry allows the choice of an experimentally invariant reference electrode. The unknown p.d. is measured by connecting the cell to a potentiometer, a source of an experimentally variable (but stable) and measurable potential difference; this p.d. is adjusted so as to reduce to zero the current in the circuit connecting the potentiometer with the cell, and then the two potential differences are equal.

---

* subsequent abbreviation p.d.

Modern potentiometers, the commonplace pH meter for example, take negligible current from the experimental cell during the process of measurement. The polarity of the probe electrode relative to the reference electrode is readily found by experiment, but the way the results are written down is a matter of convention. In this book the sign of the membrane potential is that of the outside with respect to the inside. Electrophysiologists, displaying traces of potential difference against time on oscilloscope screens, commonly use the opposite convention, that is the trace moves upwards as the intracellular probe potential goes more positive.

Some kinds of electrodes are reversible with respect to a single ionic species and therefore take up a potential with respect to the solution which is a single valued function of the concentration of that ion, for example the Ag/AgCl electrode is reversible to chloride. On the other hand, although the salt bridge of the calomel electrode is filled with a chloride solution, the electrode is not reversible to chloride, indeed its indifference to the concentration of chloride and other common biological ions at varying concentrations is a reason for its widespread use. The basis of this indifference is instructive. Clearly the p.d. between the copper connection and the KCl solution is invariant, so any cause of variation would be at the tip. Is there a potential difference at this tip? Before answering this question, suppose for a moment that the electrode were filled with concentrated NaCl solution instead of KCl. The mobility of Na^+ in solution is known to be less than that of Cl^-. Na^+ and Cl^- ions diffuse down their concentration gradient out of the electrode tip. Because the Cl^- ions tend to outstrip the Na^+ ions, a potential difference arises, the solution outside the tip going negative. The resultant electric field tends to accelerate the Na^+ ions and retard the Cl^- ions and as a result the overall process of diffusion of salt remains electroneutral. At this composition boundary, as at most, the strict electroneutrality observed in terms of chemical orders of magnitude (compare Section 3.3) inevitably results in the appearance of an electrically detectable difference of potential. The subtlety of using KCl for the salt bridge is that the mobility of K^+ and Cl^- are equal in solution, so there is no potential difference at the electrode tip. The only way in which the solution components could result in a tip potential would be if they exerted different effects on the mobility of K^+ and Cl^-, and there is little evidence for such effects.

We are concerned here with electrochemical cells in which two solution compartments are separated by a membrane. We may think of the electromotive force* of such a cell as the appropriate algebraic sum of three components, two potential differences between electrodes and the surrounding solution, and one across the membrane; while it is impossible rigorously to analyse the measured e.m.f. into these components, some reasonable guesses are in order. If the two solutions and the two electrodes were identical, clearly the measured p.d. would be that across the membrane, (a principle which underlay the description in Section 2.2 of the classical experiment with frog skin). Since the solutions would not be identical in most biological situations, let us examine the simplest analogous case, namely the Donnan membrane system (Figs. 3.10 and 5.1). If a pair of Ag/AgCl electrodes were inserted, one on each side of the system, no p.d. would be detected because the whole electrochemical system is at equilibrium and therefore $\Delta G = -FE = 0$. If on the other hand we use the 'biological' technique of inserting a pair of calomel electrodes, which are unaffected by the different chloride concentrations on the two sides, a membrane potential E_m is measured. Experimental study shows that it conforms with the Nernst equation:

$$E_m = \frac{RT}{F} \ln \frac{[Na_i]}{[Na_o]} = \frac{RT}{F} \ln \frac{[Cl_o]}{[Cl_i]} \quad (4.1)$$

*subsequent abbreviation: e.m.f.   43

This equation simply expresses the free energy condition for equilibrium (see Sections 3.3 and 5.2) in a different way. In discussing the double Donnan model for the ionic distribution between a cell and its surroundings, the argument was outlined that the same kind of condition for equilibrium could be applied to the diffusible ions which were unchanged chemically as a result of their passage across the membrane. It follows that the Nernst equation is applicable to the model in the form:

$$E_m = \frac{RT}{F} \ln \frac{[K_i]}{[K_o]} = \frac{RT}{F} \ln \frac{[Cl_o]}{[Cl_i]} \quad (4.2)$$

The left hand equation has been tested for muscle and other tissues, best of all by rapidly altering the extracellular potassium concentration $[K_o]$ and measuring E_m before any detectable change in $[K_i]$ has occurred: over a wide range of $[K_o]$ (above about 10mM) the plot of E_m against $-\ln[K_o]$ is a good straight line. An example has already been given of the validity of the right hand equation in the antilog form, and the relationship was exhaustively checked by Boyle and Conway. This might seem to be a very satisfactory outcome, and could be regarded as justifying the model. But let us restate from an electrical point of view a fundamental assumption of the model, that the sodium pump generates an outward directed electrical current (of sodium ions) through the membrane, which results in a potential difference, and that this p.d. aids the diffusion of Na^+ back into the cell and is also responsible for the distribution of K^+ and Cl^-. Seen in terms of this restatement the matter cannot be left until evidence of a different kind has been considered.

The fundamental principles of the measurement of bioelectricity have been described at some length as an aid to understanding the way potential differences arise as well as how they are measured. The biological membrane phase is charged, and it will therefore show different selectivities for the uptake of cations and anions (see Sections 3.4 and 5.2); the rate of permeation of cations and anions through the membrane under a gradient of electrochemical potential should therefore differ. Evidence of such differences of ion permeability has often been observed in electrical measurements on biological membranes. Having seen that a boundary p.d. arises from the diffusion of salt from an electrode tip when the ionic mobilities are unequal, it is apparent that similar potential differences will arise from the diffusion through a membrane of ions with different mobilities therein. This raises the question, long a subject of physiological controversy, of whether the double Donnan model for ionic distribution is the right one. In this model the ion pump is described as 'electrogenic', that is it causes the net movement of ionic charge without chemical coupling to the movement of other cations and anions, and thus makes a direct contribution to the membrane potential. An alternative model is an electroneutral pump in which the movement of one Na^+ outwards is chemically coupled with movement of one K^+ inwards. This pump creates an asymmetric distribution of cations, thereby setting up concentration differences, and therefore diffusion of cations across the membrane gives rise to a p.d. which also affects movement of anions. It is impossible to predict *a priori*, the potential difference which will result from this model, but if the steady state concentrations of ions and their permeabilities P are known from experiment, then the approximate membrane potential is given (on the strict assumption that the pump is electroneutral) by the Goldman [19] equation:

$$E_m = \frac{RT}{F} \ln \frac{P_K[K_i] + P_{Na}[Na_i] + P_{Cl}[Cl_o]}{P_K[K_o] + P_{Na}[Na_o] + P_{Cl}[Cl_i]}$$

Under normal physiological conditions in which $P_K \approx P_{Cl} > P_{Na}$ this equation gives approximately the same answer as the Nernst equation, and because of the assumptions which underlie the Goldman equation and also the fact that in either equation concentrations must perforce be used instead of activities, comparison of the experimental membrane potential with the calculated E_m does not constitute a test which will distinguish the electrogenic from the electroneutral pump model.

An alternative test is to load the cell, or tissue, with Na^+ and then observe the response of the membrane potential to the net flux situation created by returning the cell to a normal medium. It is argued (to simplify somewhat) that the electrogenic pump will give rise to an increased membrane potential in the course of shifting the Na^+ load, and in order to ensure its exchange with an increased influx of K^+, whereas the electroneutral pump would maintain a lower membrane potential, roughly that under normal conditions, determined almost entirely, in view of the low P_{Na}, by the potassium and chloride distributions. From heroic experiments on this principle, Adrian and Slayman concluded that in frog muscle in rubidium media (replacing K^+), the sodium pump was certainly electrogenic, and in normal K^+ medium there was evidence for some electrogenicity but a degree of chemical coupling of K^+ and Na^+ fluxes could not be ruled out. A more clear cut example of the comparison between observed membrane potential and expectation on the basis of the Goldman equation, as a criterion of electrogenesis, is given by Slayman [37] in a review of evidence for the presence of electrogenic proton pumps in micro-organisms.

Compelling evidence for electrogenicity of the sodium pump comes from the elegant experiments of Thomas [39] on large neurones of the common snail. Five electrodes could be simultaneously inserted into a neurone: one to measure the membrane potential, one to clamp it (i.e. to apply an electronically regulated potential bias across the membrane so as to maintain constant membrane potential while

measuring the current – compare Section 2.2), two electrodes with which to inject salt by ionophoresis, the amount being measurable electrically, and one ion-selective electrode to measure the intracellular sodium concentration. In experiments without the voltage clamp, the injection of about 25×10^{-12} equivalents of Na^+ (but not of K^+ or Li^+) with acetate anions, caused an increase of E_m (hyperpolarization) of as much as 20 mV; this response was blocked by ouabain treatment, or use of a K^+ free bathing medium. When the membrane potential was clamped, there was a rapid increase of current in the clamp circuit on injection of Na^+, followed by an exponential decrease. This current was explained if about 28% of injected Na^+ was pumped out electrogenically at a rate proportional to the excess intracellular sodium concentration. There was no evidence for variable stoichiometry.

In conclusion, p.d. measurements across biological membranes can, under favourable conditions, provide evidence about the electrochemical nature of transport mechanisms, and do currently support the assertion that the common sodium pump is electrogenic.

4.2 Water transport.

It would be misleading to write a book about biological transport which did not mention the transport of water, but to write about it at any length in so short a book would contribute no new insight to the subject, which has been reviewed with the utmost clarity: this section is therefore intended only as an aperitif to the further reading suggested below.

Water transport is of fundamental physiological importance. Large volumes of water are transported in animals across epithelial layers. Water transport is not a stoichiometric phenomenon and cannot be directly explained in terms of vectorial metabolism; it is secondary to solute transport, occurring as a result of the creation of local osmotic gradients by the active transport of solute. The epithelial layers have in common multiple membranous infoldings which enclose narrow lateral intercellular spaces. One end of each space is closed and the other is open. Mitochondria are found adjacent to the space; presumably they provide the energy for the transport of salt from the cell into the closed end of the space. The presence of this salt in the space continuously draws water from the cells, and the water and salt flow down the narrow channel to the open end. Diffusion of salt from its source at the closed end establishes a decreasing concentration gradient down the channel. The solution is nearly always isotonic when it reaches the open end and passes out of the epithelium. In summary: a major function of salt transport is to bring about the physiologically necessary movement of water.

4.3 Ion fluxes and the Na/K activated ATPase.
In this section it is shown that there is a very close correlation between the characteristics of the Na/K activated ATPase and those of the porter system responsible for the sodium pump. Much of the work described here was done on red cells by Glynn [13, 14] and his colleagues; this summary is intended to guide the reader through the original literature and not to substitute for it.

A basic technique was the measurement of unidirectional fluxes in the steady state, effluxes being measured after pre-loading by incubating the cells with tracer solution and washing them in non radioactive Ringer's solution. In a number of experiments the composition of the Ringer solution was modified by substitution of one cation for another. Once it had been shown that active fluxes and ATPase activity were inhibited by ouabain it became customary to eliminate from the interpretation of the experimental results any background activity or flux which was insensitive to ouabain by considering the difference between the normal experiment and a control in the presence of ouabain: such

differences are referred to as 'ouabain sensitive'. The reader may assume that in this section all but the earliest work (Table 4.1, fluxes 1 through 4) refers to ouabain sensitive quantities; the interesting but quantitatively minor question of ouabain insensitive fluxes is ignored.

Table 4.1. Components of ion fluxes of human red cells.

No.	Type	Glucose dependence	Characteristics
1	K influx	+	Active, saturable, $K_m \approx$ 2 mM
2	Na efflux	+	Active. Dependence on $[K_0^+]$ as flux 1, with which it is linked.
3	K influx	−	Passive, linear with $[K_0^+]$. High temperature coefficient.
4	Na efflux	−	Passive, almost independent of $[K_0^+]$.
5	Na influx	−	At high $[K_0^+]$. Passive, not directly proportional to $[Na_0^+]$, as if some dependence on efflux. High temperature coefficient.
6	Na influx	−	At low $[K_0^+]$. Markedly increased by reduction of $[K_0^+]$ as if Na^+ enters on the K carrier.
7	K efflux	−	Almost independent of $[K_0^+]$. Passive but inexplicable unless carrier postulated. High temperature coefficient.

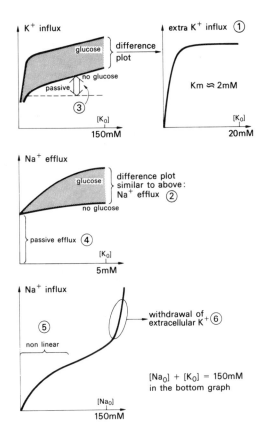

Fig. 4.1 Unidirectional ion fluxes. The numbers refer to Table 4.1. Concentration of K^+ is varied in the medium (replacing Na^+).

The results of the basic study of unidirectional potassium and sodium fluxes at varying medium concentrations of these ions are presented in terms of a composite set of sketch graphs (Fig. 4.1) and a corresponding list (Table 4.1) of the components of influx and efflux which can be recognized, together with summaries of their characteristics. By taking the differences between fluxes in glucose Ringer and those of glucose starved and washed cells in glucose-free Ringer, Glynn was able to

distinguish two fluxes, K^+ influx and Na^+ efflux (fluxes 1 and 2 in Table 4.1) which were glucose dependent. These metabolically linked fluxes are active by the criterion of Section 2.2. Both fluxes were shown to follow Michaelis Menten kinetics with respect to the extracellular potassium concentration $[K_0]$, and had the same K_m with respect of K^+. This was good evidence for their mutual mechanistic linkage. In other experiments, on the stoichiometry of these ouabain sensitive fluxes, it was found that the

47

ratio of ^{24}Na efflux to ^{42}K influx was $1\cdot20 \pm 0\cdot01$ and on another occasion $1\cdot35 \pm 0\cdot01$. It is now generally assumed that $3Na^+$ are extruded in exchange for the entry of $2K^+$, but the possibility of variable stoichiometry still has to be borne in mind. The passive K^+ influx (flux 3) shows a Fick's law relationship with $[K_0]$, but its temperature coefficient is much higher than that for the diffusion of ions in free solution. Passive Na^+ influx can be analysed into two components with different characteristics (fluxes 5 and 6). Flux 7, K^+ efflux, and flux 3 are both passive, but it appeared from a detailed consideration of their relationship that they could not be explained solely on the basis of leakage through the membrane by diffusion, but that some carrier site must be involved.

An important series of experiments was done by Whittam [43, 44] and colleagues on resealed ghosts, which have the advantage that the inside composition can be varied, and the ATPase activity can be followed by measurement of inorganic phosphate release. In the presence of sodium both inside and out, the effect of change of medium potassium concentration on ATPase activity was of the same form as that of $[K_0]$ on flux 1, the K_m for potassium being about $2\,mM$. One interesting observation, the significance of which will be noted later, was that at low $[K_0]$ in a choline medium, ATPase activity, exceeded that in a sodium medium. Changing the inside sodium concentration showed that ATP splitting was half maximal, somewhere between 60 and $80\,mM$ $[Na_i]$, and maximal above $100\,mM$. The activation of the enzyme by external cations appeared to correspond with selectivity order X (Table 3.3), whereas those of external cations followed either order III or IV. The site of release of inorganic phosphate was shown to be internal. This study therefore gave much more information about the enzyme as a porter than work on fragmented membrane preparations. The directional effects are those of a vectorial enzyme, the site specificity of which changes

according to whether the site faces inwards or outwards. The ouabain sensitive activation by Na^+ and K^+ is such that it suggests that the ions activate the enzyme in those locations from which flux studies show that they are actively transported. The 'energy consuming' enzyme seems to have the same vectorial character as the porter, and one assumes that those ions which activate are also transported, though this is not actually proved by the experiments.

We next examine the relationship between the vectorial enzyme of red cells and the Na/K activated ATPase found in other tissues, the possible vectorial properties of which are difficult to study experimentally. Extensive comparative studies [2] of this ATPase activity in various tissues and species with the corresponding rates of sodium pumping show a good correlation, the range of rates being very wide.

Another set of comparisons depends on more detailed biochemical study of the isolated enzyme on the lines of experiments first done on the kinetics of ATP splitting in a preparation from guinea pig kidney [36]. It was shown, by working at low temperatures to slow down the reaction, that a phosphorylated intermediate is rapidly and reversibly formed; its formation requires sodium and magnesium, but there was some uncertainty about the sequence of events, and its breakdown requires potassium. The level of phosphorylated intermediate was measured using a labelling technique. Under standard conditions the turnover (in units of time^{-1}) was found as the ratio of ATPase activity to the level of the phosphorylated intermediate [1]. This turnover, derived from measurements in different tissues and species with a wide range of enzyme activity, varied over only a roughly two-fold range. The near constancy of this parameter is an indication that species differences in the properties of the enzyme are small. This awaits confirmation from the study of protein characteristics, but the purification of the enzymes presents problems.

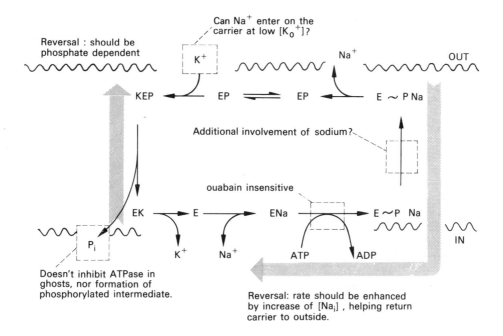

Fig. 4.2 Modification of Shaw's model in the light of results on phosphorylated intermediate and internal release of inorganic phosphate. Broad arrows indicate possible modes of reversal.

A further outcome of studies on the guinea pig enzyme is a generalization about the factors affecting the rate of ATP splitting. Conditions were found in which experimentally variable concentrations could be made rate limiting, and the level of the phosphorylated intermediate was measured. Its amount was found to vary as the reaction rate when the concentrations of ATP, Mg^{++} or Na^+ were rate limiting, but to vary inversely as the rate when the K^+ concentration was limiting. The implications of these and other findings are summarized diagrammatically in Fig. 4.2 which is basically the carrier scheme originally proposed by Shaw as an interpretation of the fluxes, modified to account for various experimental findings up to about 1967. Fig. 4.2 also hints at characteristics of partial reversal of the sodium pump if this carrier scheme were valid.

Overall reversal of the pump with concomitant synthesis of ATP was but one of the numerous important results published in 1967 in a set of papers by Garrahan and Glynn on experiments in which cells and ghosts were submitted to non-physiological conditions in order to further probe the mechanism of ion transport. With cells in a potassium free medium, a one for one exchange of intracellular and extracellular sodium was shown which was proportional to $[Na_0]$ from high concentrations down to 5 mM. This dependence on $[Na_0]$ at concentrations generally greater than $[Na_i]$ could not be explained as 'exchange diffusion' mediated by a system such as that shown in Fig. 3.8 (which should have the same K_m for sodium inside as outside); it was therefore necessary to postulate a change of conformation and affinity on crossing the membrane, indicated in Fig. 4.3

49

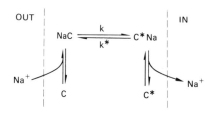

OUT | | IN

NaC $\underset{k^*}{\overset{k}{\rightleftharpoons}}$ C*Na

Na^+ — C   C* — Na^+

Fig. 4.3

by the change of the carrier site from the form C to C*, for which the ratio of rate constants k and k* would not be equal to 1. Something of this kind might be expected if the sodium exchange were occurring in the part of the cyclic pathway which was energy coupled. By studying this exchange in ghosts and cells under abnormal conditions it was found that when the inside sodium and ATP concentrations were reduced to low levels and the inside inorganic phosphate concentration was raised, the $Na^+:Na^+$ exchange (in K free medium) exceeded the coupled $Na^+:K^+$ exchange under normal conditions. The abnormal conditions stated are such as to make the free energy available for sodium pumping less negative, or perhaps even positive which should bring about reversal of the sodium pump. The reader should note the paradoxical increase of sodium efflux by reduction of the internal sodium concentration.

Overall reversal of the pump was shown in ghosts with appropriately adjusted internal and external concentrations of Na^+, K^+, ATP, ADP and P_i as a significant incorporation of ^{32}P into ATP in the presence of iodoacetate, an inhibitor of glycolysis. In other experiments it has been shown that the cyclic pathway can be reversed as a whole or in part by the correct choice of conditions.

A cyclic scheme which is consistent in detail with all the evidence briefly reviewed here is that of Stone [38]; a diagram of the main

features of his model, together with some comments, is shown in Fig. 4.4. The initial basis of this model was a computer simulation study of the kinetic characteristics of the elements of possible carrier schemes for ion transport, in comparison with the kinetics of sodium fluxes and other relevant data already reported in the literature for a variety of tissues. Schemes in which the translocation steps were irreversible (physiologically unidirectional) could be eliminated from further consideration, and schemes in which the interaction of Na^+ and K^+ with a carrier site was sequential were preferred to those in which they interacted simultaneously. In the final model the intermediates in the groups of reactions enclosed by dotted lines are close to equilibrium at all times (compare Section 3.4). The parameters of the model were adjusted so that it was quantitatively consistent with all the suitable data on red cell and kidney systems, with one exception which was qualitatively of the right form. The model became unsatisfactory with regard to the phosphorylation kinetics, if phosphorylation of E was assumed to precede sodium binding. The model explains two particular points which have been mentioned above, the paradoxical effect of reduction of $[Na_i]$ on the $Na^+: Na^+$ exchange (reversal of step 1) and the difference between the effects of outside sodium and choline in experiments on ghosts in low potassium medium (step 2). The model transports charge outwards, and is therefore electrogenic, and the overall charge of the carrier site (larger cycle) changes when it is phosphorylated. The ouabain insensitive phosphorylation (of *E*) is in the subsidiary cycle. In conclusion this model, which Stone was very careful to describe as hypothetical, does seem to be a very good one.

Bibliography
Moore, W.J. (1972) *Physical Chemistry*, 5th edition. Longmans, London.
Chapter 12 takes a fundamental approach to electrochemistry.

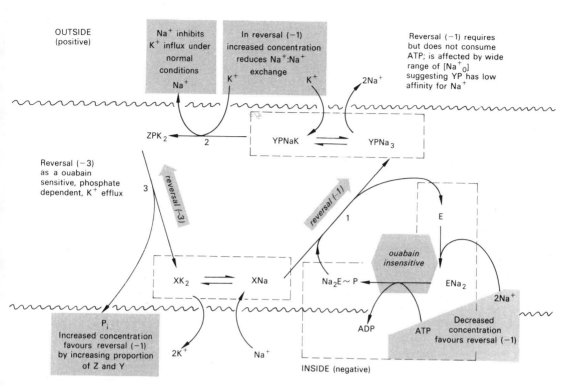

OUTSIDE
(positive)

Na⁺ inhibits K⁺ influx under normal conditions Na⁺

In reversal (−1) increased concentration reduces Na⁺:Na⁺ exchange

Reversal (−1) requires but does not consume ATP; is affected by wide range of $[Na^+_0]$ suggesting YP has low affinity for Na^+

Reversal (−3) as a ouabain sensitive, phosphate dependent, K⁺ efflux

ouabain insensitive

P_i Increased concentration favours reversal (−1) by increasing proportion of Z and Y

Decreased concentration favours reversal (−1)

INSIDE (negative)

Fig. 4.4 Composite diagram of Stone's model for the sodium pump, together with comments illustrating certain features.

Diamond, J. (1965) *Symp. Soc. Exp. Biol.* **19** 329 and (1971) *Phil. Trans. Roy. Soc.* **B 262** 141.
Lucid exposition of experiments on water transport, and their implications.
Gutknecht, J. (1970), *Am. Zoologist,* **10**, 347.
A concise explanatory review on the origin of bioelectric potentials in plant and animal cells.

5 Appendices

5.1 Capsule thermodynamics: a reminder.

The word *system* is often used in thermodynamics, and sometimes carries a thermodynamic connotation in this book. A system is an assembly of interdependent parts within a boundary, the boundary being a mental device defining that part of the universe in which we are interested. The boundary may correspond with some real physical container. Physicists often speak of idealized systems such as frictionless pulleys in order to help understand a basic principle. Biologists follow their example and construct model systems. Sometimes they do not bother to define the boundary too carefully because it can be a bore to be too pedantic: however such vagueness can also be hazardous. (Sometimes they refer to a *black box*. This is a boundary which contains an unknown mechanism.) An *open* system is one which can exchange energy and matter with its surroundings: living cells are open systems. When a cell is in a *steady state* we mean that its input (of energy and matter) is equal to its output; this carries the implication that the concentration of all the intermediates between the input substances (fuels) and output substances (waste products and secretions) are constant.

We know that *energy* is conserved (First Law) though the zero of the energy scale is arbitrary. We remember that energy can be interconverted amongst various forms, but tend to forget that these forms are really different ways of expressing measurements of the effects of motion or components of motion of the ultimate particles

of matter. There is a rule about the interconversion of heat and work (Second Law) which is generalized from experience (for example we notice that heat is evolved if a car battery is rapidly discharged, but we cannot recharge the battery by heating it). Heat is the most degraded 'form' of energy. The change in *enthalpy* ΔH in a reaction is the heat change under defined conditions. The change in *entropy* ΔS in a reaction is the change in the way heat energy is spread amongst the different energy levels which are, in order of decreasing magnitude: electronic, vibrational, rotational and translational. Often molecules of a substance at room temperature must all have the same electronic configuration (one energy level) but may be found in many different translational levels. In a reaction in which the products can, so to speak, choose between more energy levels than the reactants (ΔS positive) this factor tends to favour product formation; this tendency may however be counteracted if the reaction occurs with the absorption of heat (ΔH positive). The balance of these tendencies is expressed by the *free energy* change (ΔG) which is given by the equation $\Delta G = \Delta H - T\Delta S$, in which T is the absolute temperature. It is a mistake to try and visualize free energy, or use analogies; it is a composite thermodynamic function of two components of different character. Certainly it should not be confused with heat, something which is absorbed or evolved. The free energy G of a system has the important property that it is at a minimum when the system is an equilibrium; as systems approach equilibrium their free energy therefore decreases. Moving towards equilibrium is a *spontaneous process,* in which ΔG is therefore negative. Such a spontaneous process may be made to do work against its surroundings, but in any real process this work must be less than ΔG (Second Law again). Conversely work has to be done on this system in order to increase ΔG: looked at from the point of view of the system this would not be a spontaneous change. However all natural processes are spontaneous overall (note in passing that this statement depends on the way in which we define the boundaries). The free energy of a component of a system (e.g. water in a solution) depends on the composition of the system: the *chemical potential* μ is the free energy per mole (defined in a special way which need not concern us here). The chemical potential is related to the *activity a,* which is the concentration multiplied by a coefficient such that the equation $\mu = \mu^0 + RT \ln a$ is obeyed. The term μ^0 is the standard chemical potential under agreed reference conditions. The standard free energy changes ΔG^0 for reactions are defined in a similar way, the reference conditions being at one atmosphere pressure and 25°C for the pure substance, or a solution at unit activity. If the concepts worry you, remember the operational convenience of ΔG^0, which is related to the equilibrium constant K by the equation $\Delta G^0 = -RT \ln K$. Also there is an equation for finding ΔG for a reaction occurring at specified concentrations provided that ΔG^0 is known. Reactions in which movement of electric charge takes place may be studied by setting up the corresponding electrochemical cells: the electromotive force E and free energy change are related by the equation $\Delta G = -nFE$, where F is the Faraday and n the number of electrons transferred. Diffusion from a high concentration, say $[S'']$ to a low concentration $[S']$ is a spontaneous process, for which the free energy change is $RT \ln [S']/[S'']$; conversely the free energy required for the process of osmotic concentration from $[S']$ to $[S'']$ is $RT \ln [S'']/[S']$.

Free energies are very convenient because they can be added whenever it is permissible to add the corresponding reactions, as in sequences of reactions with common intermediates. When the existence of a common intermediate allows us to add them, we speak of a pair of reactions as *coupled.* If we are ignorant of the details of a biological process, tables of ΔG^0 values for biochemical reactions may help us understand which hypothetical mechanisms may be coupled with one another so that the overall process would be spontaneous (ΔG negative). Thermodynamics does not tell us that a process with a negative ΔG will necessarily occur at a detectable rate but, as biochemists, we tend optimistically to assume that nature will have provided an enzyme which catalyses the reaction which interests us. It is often stated that much of the free energy of glucose combustion is conserved as free energy of hydrolysis of ATP. The high 'efficiency' with which living cells conserve free energy really depends on the existence of very efficient catalysts which allow reactions to proceed at useful rates; it also happens that their ΔGs are rather low. Although free energies are operationally convenient for biochemical book-keeping, the typical biological steady state is more properly treated by the thermodynamics of *irreversible processes* (spontaneous processes) in which the forces operating and the resultant flows are related to the rate of entropy production. The efficiency of cells depends on their ability to minimize their rate of entropy production and maximize their state of order (or non-randomness), and thereby conserve free energy.

5.2 The Donnan equilibrium.

The membrane in the two-sided chamber shown in Fig. 5.1 is permeable to salts and water, but not to protein P. Consider setting up an equilibrium by conceptual stages. On the addition of

53

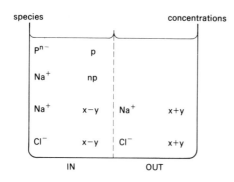

species			concentrations
P^{n-}	p		
Na^+	np		
Na^+	$x-y$	Na^+	$x+y$
Cl^-	$x-y$	Cl^-	$x+y$

IN   OUT

Fig. 5.1

sodium proteinate on one side to a solution of NaCl at concentration x, a salt shift occurs, represented by y. The salt shift is instantaneous, but may be thought of in stages. On addition of the electroneutral sodium proteinate there is an excess of sodium on the left (newly added sodium ions being indistinguishable from the rest) as a result of which some sodium ions migrate to the right. The requirement for strict electroneutrality enforces an equal movement of chloride. The net ion movement ceases when the condition for equilibrium is met. The symbols used here, and the definition of concentrations of the diffusible ions as sums and differences is not the usual one, but has the advantage of avoiding a quadratic equation*. Provided the correct pressure $\Delta\pi$ is exerted across the membrane (inside positive) the system comes to equilibrium. The conditions for equilibrium are: i. Electroneutrality, which is satisfied by the symbols as defined; and ii. Free Energy: the chemical potential of salt on the two sides must be equal. It follows from condition ii. that:

$$\frac{[Cl_o]}{[Cl_i]} = \frac{[Na_i]}{[Na_o]}$$

* I am indebted to Dr A.G. Ogston for the clever symbols and definitions. His approach to the Donnan equilibrium is published here for the first time.

54

Substituting the quantities defined by the diagram:

$$(x + y)^2 = (x - y + np)(x - y)$$
$$\therefore \cancel{x^2} + 2xy + \cancel{y^2} = x^2 - 2xy + npx - npy + \cancel{y^2}$$

Solving for y:
$$y = \frac{npx}{np + 4x}$$

The osmotic pressure difference is given by:

$$\Delta\pi = RT(p + np - 4y)$$

Finally, note that a homogeneous membrane which bears fixed charges, is ion selective because it excludes some of the co-ions (of the same polarity as its fixed charge) on a Donnan basis.

5.3 The sodium pump and the double Donnan distribution.

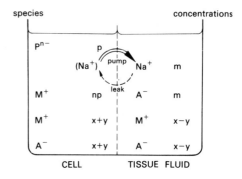

species			concentrations
P^{n-}	p		
	(Na$^+$) pump Na$^+$		m
M^+	np leak A^-		m
M^+	$x+y$	M^+	$x-y$
A^-	$x+y$	A^-	$x-y$

CELL   TISSUE FLUID

Fig. 5.2

This is a formal scheme, postulating a single pumped ion, Na$^+$, which is different from other univalent cations represented by M$^+$. To 'create' such a pump we have to provide an equivalent concentration of A$^-$ in the tissue field to preserve electroneutrality. (We could just as well have started with Na$^+$ and A$^-$ equally distributed and let Na$^+$ be pumped, some A$^-$ following passively). We do not assume that the mechanism of the sodium pump necessitates reverse potassium transfer for any mechanistic reason, though in fact this assumption would not alter what follows provided M^+ includes enough K^+. In practice there is Na^+ inside which has leaked in, but we boldly treat leaked Na^+ as if identical with M^+. The symbol P^{n-} here represents the functional 'fixed negative charge', with mean negative charge n. The direction of the salt shift y cannot be predicted in general.

Condition ii. of the Donnan equilibrium (Section 5.2) applies to M^+ and A^-, although the present system is a steady state in which the rate of pumping of Na^+ equals its rate of leakage back. From condition ii.:

$$(np + x + y)(x + y) = (x - y)(x - y + m)$$

$$\therefore 4xy + npy + my = mx - npx$$

$$\therefore y = \frac{mx - npx}{4x + np + m}$$

$$\therefore \Delta\pi \propto (p + np - 2m + 4y)$$

The significant point is that right regulation of the pump can lead to $\Delta\pi = 0$, which corresponds with nearly all experimental evidence for mammalian cells. We identify M^+ outside as K^+, and A^- outside as chloride and bicarbonate. By substituting values from Table 2.2 we reach the conclusion that to a first approximation $y = 0$. The membrane potential is given by the Nernst equation:

$$E_m = \frac{RT}{F} \ln \frac{[M_i]}{[M_0]} = 1 + \frac{m + np}{2x}$$

Experimentally $E_m = \dfrac{RT}{F} \ln \dfrac{[K_i]}{[K_0]} = \dfrac{RT}{F} \ln \dfrac{[Cl_0]}{[Cl_i]}$

The sign convention here is that of the 'outside', with respect to the 'inside' as zero. Note how changes of m, np, and x will change E_m, and compare with experimental results reported in the literature.

Bibliography

Donnan, F.G. (1925) *Chem. Rev.* **1**, 73.
His review of his equilibrium.
Linford, J.H. (1966) *An introduction to energetics with application to biology.* Butterworths, London.
For the student in the armchair who appreciates mastery of style and content.
Morris, J.G. (1968) *A biologist's physical chemistry.*
The book for the student in the laboratory, with a slide rule in his pocket.

Topics for further reading

The arguments presented in this book have deliberately been illustrated with a restricted range of biological examples which mainly concern mammalian red cells and muscle, and occasionally bacteria, frog skin, and nerve. There may be readers who wish to relate the principles explained in this book to other species and tissues in which they are interested. This section therefore consists of a list of references and comments intended to provide 'seed thoughts' on a variety of topics. No doubt the topics chosen will not satisfy all tastes, but further interesting studies of biological transport and membrane function are likely to be revealed by perusal of the tables of contents and indexes of the following publications: Physiological Reviews, Annual Review of Physiology, Annual Review of Biochemistry, Annual Review of Plant Physiology, Federation Proceedings (symposium issues), British Medical Bulletin, Harvey Lectures, Quarterly Reviews of Biophysics, Advances in Botanical Research, Progress in Biophysics and Molecular Biology.

The references listed below are intentionally not always the latest, and will in any case go out of date within a short space of time. Fortunately it is easy to find up-to-date references on any scientific topic by using Science Citation Index, which consists of computer analyses (the Citation Index and the Permuterm Index) of the bibliographies of all scientific papers published in an enormous range of journals, and a check list of all the titles of all the papers analysed, alphabetically by all authors (the Source Index): Science Citation Index cumulates annually. To update one of the topic lists below, find out from the Citation Index of the current year who has quoted one, or better two, of the papers which interest you. If your two choices have been quoted in the same paper, it is very likely that its author has linked these papers because of a similar interest to your own: however, it may be wise to check the titles in the Source Index. This method of literature searching tends to be very specific, so if you find nothing, it probably means there really is no news. If you want to broaden the scope of your inquiry, try looking up in the Permuterm Index, the users in paper titles of indicative word pairs (e.g. yeast/membrane, and yeast/phosphate, but not membrane/transport): full titles will be found in the Source Index. This method of searching should prove much more rapid, specific and complete than the use of conventional abstracts. (A magnifying glass may prove useful.)

Protozoan contractile vacuole
How animals without kidneys get rid of excess water.

Kitching, J.A. (1938) *Biol. Rev.* **13** 403.
Kitching, J.A. (1954) *Symp. Soc. Exp. Biol.* **8** 63.
Kitching, J.A. (1967) in T.T. Chen (ed) *Research in Protozoology* Vol 1, 309, Pergamon Press, Oxford.
Schmidt-Nielson, B. and Schrauger, C.R. (1963) *Science* **169** 606. Micropuncture studies.
Prusch, R.D. and Dunham, P.B. (1970) *J. Cell. Biol*; **46** 431.

Gastric secretion
Rehm, W.S. (1972) *Arch. Intern. Med.* **129** 270. The location and mechanism of proton secretion. Forte, W.G. (1970) in Bittar, E.C. *Membranes and ion transport,* Vol 3, p 111. Wiley-Interscience, New York. A general review.

Ion transport in the nervous system
All but the last two references are general, and in some cases quite elementary, accounts of different aspects. The first five were special lectures.
Hodgkin, A.L. (1958) *Proc. Roy. Soc.* **B 148** 1.
Eccles, J.C. (1964) *Science* **145** 1140.
Hodgkin, A.L. (1964) *Science* **145** 1148.
Huxley, A.F. (1964) *Science* **145** 1154.
Katz, B. (1971) *Science* **173** 123.
Eccles, J.C. (1965) *Scientific American,* January, 56.
Baker, P.F. (1966) *Scientific American,* March, 74.
Carpenter, D.O. and Alving, B.O. (1968) *J. Gen. Physiol.* **52** 1 Metabolic dependence of membrane potential.
Gorman, A.L.F. and Marmor, M.F. (1970) *J. Physiol.* **210** 879, 919. Another study of similar phenomena.

Potassium requirement for protein synthesis
Perhaps as relevant to ion transport as the geochemical factors mentioned in Section 3.2
Pestka, S. (1970) in Bittar, E.E. (ed) *Membranes and ion transport,* Vol 3, p 279, Wiley-Interscience, New York.

Isolation of transport proteins from bacteria
Physiological studies of transport and its inhibition, together with biochemical genetics, defined the existence of 'permeases' on which the uptake of substrates by bacteria depends. More recently proteins implicated in transport systems have been isolated and in some cases crystallized.
Cohen, G.N. and Monod, J. (1957) *Bact. Rev.* **21** 169. Classical review on permeases.
Fox, C.F. and Kenedy, E.P. (1965) *Proc. Nat. Acad. Sci.* **54** 891. *Lac* M protein of E. coli.
Fox, C.F. et al (1967) *Proc. Nat. Acad. Sci.* **57** 698. *Lac* M protein of E. coli.
Pardee, A.B. and Prestidge, L.S. (1966) *Proc. Nat. Acad. Sci.* **55** 189. Sulphate binding protein.
Pardee, A.B. (1968) *J. Gen. Physiol.* **52** Supplement 279. Review of biochemical studies.
Kepes, A. (1971) *J. Membrane Biol.* **4** 87. The β-galactoside permease reviewed.

Osmotic shock and the phosphotransferase system of bacteria
Components of the PEP linked system and other enzymes and proteins which bind various substrates are released into the medium by suddenly decreasing its osmolarity. The obligatory linkage of phosphorylation with transport (see Section 3.1.1) has been shown in experiments on closed vesicles prepared from bacteria.
Heppel, L.A. (1967) *Science* **156** 1451.
Simoni, R.D. et al (1967) *Proc. Nat. Acad. Sci.* **58** 1963.
Kaback, H.R. (1970) in Bronner, F. and Kleinzeller, A. (eds), *Current Topics in Membranes and Transport,* Vol 1, p 34, Academic Press, New York.

Lipid bilayers and ionophores
The properties of artificial lipid bilayers resemble those of natural membranes. The ion permeability of lipid bilayers set up *in vitro* between two compartments is increased by cyclic compounds, including certain antibiotics, which bind an ion at a site enclosed within a hydrophobic 'wrapping': these ionophores appear to resemble transport sites.
Henn, F.A. et al (1967) *J. Mol. Biol.* **24** 51.
Thompson, T.E. and Huang C. (1966) *Ann. N. Y. Acad. Sci.* **137** (2) 740.
Mueller, P. and Rudin, D.O. (1967) *Biochem. Biophys. Res. Comm.* **26** 398.
Kilbourn, B.T. et al (1967) *J. Mol. Biol.* **30** 559.

Krasne, S. et al (1971) *Science* **174** 412.
Harold, F.M. (1971) in Rose, A.F. and Wilkinson, J.F. *Advances in Microbial Physiology* **4** 45, Academic Press, London.

Metabolite transport in mitochondria
Chapell, J.B. (1968) *Brit. Med. Bull.* **24** 150.
Tubbs, P.K. and Garland, P.B. (1968) *Brit. Med. Bull.* **24** 158, in which Fig. 2 is a good example of a vectorial system.
Klingenberg, M. (1970) in Campbell, P.N. and Dickens, F. (eds) *Essays in Biochemistry* **6** 119, Academic Press, London.
McGivan, J.D. and Chappell, J.B. (1972) *Biochem. J.* **127** 54P.

Ion transport in algae and plants
Sutcliffe, J.F. (1962) *Mineral Salts absorption in plants* Pergamon, Oxford.
Briggs, G.E., Hope, A.B. and Robertson, R.N. (1961) *Electrolytes and plant cells* Davis, Philadelphia.
MacRobbie, E.A. (1970) *Q. Rev. Biophys.* **3** 251,
Nobel, P.S. (1970) *Plant cell physiology* Freeman, San Francisco.
Saddler, H.D.W. (1970) *J. Gen. Physiol.* **55** 802.

The Chemiosmotic Hypothesis
According to an important hypothesis of Mitchell, the coupling of oxidation and phosphorylation in the production of ATP (for example in mitochondria and chloroplasts) is mediated by the gradients of proton and other ion concentrations between the inside and outside of the organelle, that is, some of the energy of the primary vectorial process of H^+ secretion is coupled with the production of ATP. Since the hypothesis has been altered over the years and is still under detailed investigation, it did not seem to be a sufficiently immovable rock to serve as a foundation for the present book. However, care has been taken to make Chapter 3 (particularly Section 3.1) consistent with the hypothesis as understood at the time of writing.

The references chosen here are explanatory and exemplary rather than definitive.
Greville, G.D. (1969) in Sanadi D.R. (ed) *Current Topics in Bioenergetics,* Vol 3, p 1, Academic Press, New York.
Deamer, D.W. (1969) *J. Chem. Ed.* **46** 198.
Glynn, I.M. (1967) *Nature* **216** 1318.
Hopfer, U. et al (1968) *Proc. Nat. Acad. Sci.* **59** 484.

References

1. Bader, H. and Post, R.L. (1968), *Biochim. Biophys. Acta.* **150**, 41.
2. Bonting, S.L. (1970) in Bittar E.E. *Membranes and ion transport.* Vol. 1. p. 257 Wiley-Interscience, New York.
3. Boyle, P.J. and Conway, E.J. (1941) *J. Physiol.* **100**, 1.
4. Bretscher, M.S. (1971), *J. Mol. Biol.* **59**, 351.
5. Bretscher, M.S. (1972), *Nature,* **236**, 11.
6. Burton, A.C. (1941), *J. Cell. Comp. Physiol.* **14**, 327.
7. Clifford, J., et al (1968) in Bolis, L. and Pethica, B.A. (eds), p. 19 *Membrane Models and the Formation of biological membranes.* North-Holland, Netherlands.
8. Conway, E.J. and Hingerty, D. (1946), *Biochem. J.* **40**, 561.
9. Davies, R.E. and Ogston, A.G. (1950), *Biochem. J.* **46**, 324.
10. Davson, H. and Danielli, J.F. (1952), *The Permeability of Natural Membranes,* Cambridge.
11. Dawes, E.A. (1972), *Quantitative problems in biochemistry* 5th ed. pp. 167, 184 and 403. Churchill Livingstone, Edinburgh.
12. Fox, C.F. et al (1971), *Biochem. Biophys. Res. Comm.* **44**, 497 and 503.
13. Glynn, I.M. (1968), *Brit, Med. Bull.* **24**, 165.
14. Glynn, I.M. et al (1971), *Phil. Trans. Roy. Soc.* **B 262**, 91.
15. Guggenheim, E.A. (1959), *Thermodynamics* 4th ed. 372–373. North-Holland, Amsterdam.
16. Hitchcock, D.I. (1945), in Höber, R. (ed.) *Physical Chemistry of Cells and Tissues.* p. 13 Churchill, London.
17. Hoare, D.G. (1972), *Biochem. J.* **127**, 62P.
18. Hoare, D.G. (1972), *J. Physiol.* **221**, 311.
19. Hodgkin, A.L. and Katz, B. (1949), *J. Physiol.* **108**, 37 and 74.
20. Hodgkin, A.L. et al (1952), *J. Physiol.* **121**, 424.
21. Hodgkin, A.L. and Horowicz, P. (1959), *J. Physiol.* **148**, 127.
22. Kundig, W. et al (1966), *J. Biol. Chem.* **241**, 3243.
23. Leaf, A. (1956), *Biochem. J.* **62**, 241.
24. LeFevre, P.G. (1961), *Pharmacol. Rev.* **13**, 39.
25. Lehman, R.C. and Pollard, E. (1965), *Biophys. J.* **5**, 109.
26. Mitchell, P. (1961) in Kleinzeller, A. and Kotyk (A) *Membrane Transport and Metabolism,* p. 22, Academic Press, London.
27. O'Brien, J.S. (1967), *J. Theor. Biol.* **15**, 307.
28. Overath, P. et al (1971), *Nature,* **234**, 264.
29. Parsons, D.F. (1971) in Chance B. et al (Editors), *Probes of Structure and Function of Macromolecules and Membranes,* Vol. 1, page 197. Academic Press, New York.
30. Quastel, J.H. (1926), *Biochem. J.* **20**, 166.
31. Reid, E.W. (1892), *Brit. Med. J.* 1133.
32. Rosenberg, T. (1954), *Symp. Soc. Exp. Biol.* **8**, 27.
33. Rosenberg, T. and Wilbrandt, W. (1957), *J. Gen. Physiol.* **41**, 289.
34. Schultz, S.G. et al (1963), *J. Gen. Physiol.* **47**, 329.
35. Schultz, S.G. and Curran, P.F. (1970), *Physiol. Rev.* **50**, 637,

36. Sen, A.K. et al (1965), *J. Biol. Chem.* **240,** 1437.

37. Slayman, C.L. (1970), *Am. Zool.* **10,** 377.

38. Stone, A.J. (1968), *Biochem. Biophys. Acta.* **150,** 578.

39. Thomas, R.C. (1969), *J. Physiol.* **201,** 495.

40. Van. Deenen, L.L.M. (1966), *Progr. Chem. Fats. Lipids,* **8,** 1.

41. Ways, P. and Hanahan, D.J. (1964), *J. Lipid Res.* **5,** 318.

42. West, I.C. and Mitchell, P. (1972), *Biochem. J.* **127,** 56P.

43. Whittam, R.H. (1962), *Biochem. J.* **84,** 110.

44. Whittam, R.H. and Ager, M. (1964), *Biochem. J.* **93,** 337.

Index